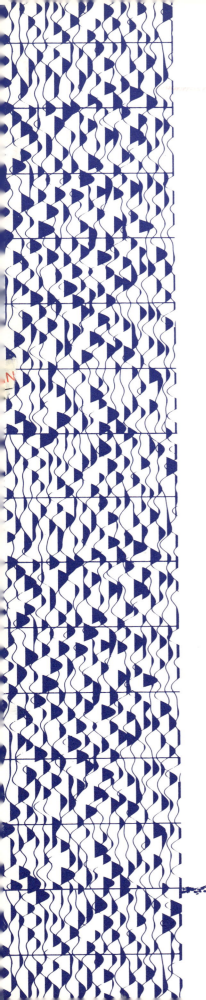

MODERN AND ANCIENT SHELF CLASTICS:

A CORE WORKSHOP

SEPM CORE WORKSHOP NO. 9

MODERN AND ANCIENT SHELF CLASTICS:

A CORE WORKSHOP

Organized and Edited
by

Thomas F. Moslow
and
Eugene G. Rhodes

SEPM CORE WORKSHOP NO. 9
ATLANTA, JUNE 15, 1986

TABLE OF CONTENTS

Page

LIST OF CONTRIBUTORS

KATHERINE M. BERGMAN
Department of Geology
McMaster University
Hamilton, Ontario
Canada L8S 4M1

ROBERT S. BRACKETT
Shell Offshore, Inc.
New Orleans, Louisiana

DAVID M. BUSH
Department of Geology
Duke University
Durham, North Carolina   27706

CLARENCE V. CAMPBELL
RPI Canada Ltd.
1616, 540 – 5th Avenue S.W.
Calgary, Alberta
Canada   T2P 0M2

W. J. EBANKS, JR.
ARCO Exploration and Production Research
2300 West Plano Parkway
Plano, Texas   75075

JOHN C. HORNE
RPI International
2845 Wilderness Place
Boulder, Colorado 80301

ROWDY C. LEMOINE
Basin Research Institute and
 Louisiana Geological Survey
Louisiana State University
Baton Rouge, Louisiana   70803

PHILIP LOWRY
Basin Research Institute and
 Louisiana Geological Survey
Louisiana State University
Baton Rouge, Louisiana   70803

LIST OF CONTRIBUTORS (Cont.)

THOMAS F. MOSLOW
Louisiana Geological Survey
Coastal Geology Program
University Station, Box G
Baton Rouge, Louisiana 70893

C. R. OSSIAN
ARCO Oil and Gas Company
P.O. Box 2819
Dallas, Texas 75221

SHEA PENLAND
Louisiana Geological Survey
Coastal Geology Program
University Station, Box G
Baton Rouge, Louisiana 70893

VISHNU RANGANATHAN
Department of Geology
Louisiana State University
Baton Rouge, Louisiana 70803

JAMES M. RINE
NL Erco/NL Industries, Inc.
225 West Airtex Drive
Houston, Texas 77090

JOHN H. RISTOW
Sedimentology, Inc.
Boulder, Colorado 80302

A. J. SCOTT
RPI Texas
5910 Courtyard Drive
Austin, Texas 78731

CHARLES T. SIEMERS
Sedimentology, Inc.
Boulder, Colorado 80302

WILLIAM L. STUBBLEFIELD
National Oceanic and Atmospheric Administration
Undersea Research Group
Rockville, Maryland 20852

JOHN R. SUTER
Louisiana Geological Survey
Coastal Geology Program
University Station, Box G
Baton Rouge, Louisiana 70893

DONALD J. P. SWIFT
ARCO Exploration Technology
2300 West Plano Parkway
Plano, Texas 75075

S. L. THOMPSON
ARCO Oil and Gas Company
P.O. Box 1610
Midland, Texas 79702

RODERICK W. TILLMAN, Consultant
4555 South Harvard Avenue
Tulsa, Oklahoma 74135

ROBERT S. TYE
Coastal Studies Institute
Louisiana State University
Baton Rouge, Louisiana 70803

ROGER G. WALKER
Department of Geology
McMaster University
Hamilton, Ontario
Canada L8S 4M1

PREFACE

Early in the design of this project we opted for a broad, permissive definition of shelf clastics, and consequently, contributors have been encouraged to focus on their own interpretations rather than force-fitting their examples to a restrictive theme of our choosing. As a result, we have assembled a selection of core examples which range from standline to shelf edge.

In the past, there has been much discussion concerning the appropriateness of including transgressed relict sediments within the context of clastic shelf depositional systems. However, we felt it was in the best interest of the workshop to avoid, where possible, any restrictions or caveats associated with the issue of relict or palimpsest shelf sediments. For the most part, we were successful. Certainly, the contention concerning the origin of, and processes active in, both modern and ancient shelf settings will continue, and this workshop presents several more strands of evidence ranging in age from the Triassic to the Holocene.

Given the dominance of petrophysical logs in subsurface studies, contributors to the workshop have worked hard to make the necessary core-to-log comparisons which are essential if geologists are to more fully utilize wireline data in facies analysis. Clearly, more geologists would be concerned with the shape and form of log response in addition to the numerical techniques which for decades have been the domain of the log analyst. The inference of lithology from logs through both qualitative and quantitative strategies offers considerable promise for the subsurface geoscientist.

We have encouraged contributors to place their studies within the context of energy prospects associated with environments of deposition. Obviously, hydrocarbon potential and reservoir quality dominate these discussions. The high degree of lithofacies variability and the effect of high-energy events on shelf deposits cause sediments formed within this environment to be particularly challenging exploration targets. The dominance of facies or stratigraphically controlled reservoirs within shelf environments suggests that continued research is required to identify the primary and diagenetic processes which control reservoir geometry in shelf settings.

During a year when economic uncertainty has imposed havoc in the energy industry, we are pleased that all of our contributors have successfully forged ahead within such a climate, many at their own financial and professional expense. Similarly, we found that the austerity measures arising from the energy sector during 1985 precluded the attendance and exhibition of all of our overseas contributors. We have great hope that future workshops will develop under conditions more favorable to the geoscience professions.

The Editors

ACKNOWLEDGMENTS

This core workshop and book were made possible only with the invaluable assistance of many people. Of foremost importance in the preparation of the course notebook was the monumental effort of the skilled and professional staff at the Columbia (South Carolina) office of Research Planning Institute, Inc. The typing, word processing, graphics, and editorial services of Cindi Fehrs, Starnell Williams, Cindy Price, and especially Linda Rader are gratefully acknowledged. Dr. Miles O. Hayes, President of RPI, Inc., is thanked for making his staff available to us.

Grammatical editing was performed by Mary Penland of the Louisiana Geological Survey. Her dedication to professionalism and detail is gratefully acknowledged. The SEPM staff supported the theme and organization of this workshop and handled logistical preparations. Donna Schemel and Joni Merkel of SEPM headquarters are especially thanked.

Maxine and Anthea are personally thanked by the editors for their patience and support.

LITHOSTRATIGRAPHY OF HOLOCENE SAND RIDGES FROM THE NEARSHORE AND MIDDLE CONTINENTAL SHELF OF NEW JERSEY, U.S.A.

James M. Rine,[1] Roderick W. Tillman,[2] William L. Stubblefield,[3] and Donald J. P. Swift[4]

[1]NL Erco/NL Industries, Inc., 225 West Airtex Drive, Houston, Texas 77090

[2]Consultant, 4555 South Harvard Avenue, Tulsa, Oklahoma 74135

[3]National Oceanic and Atmospheric Administration, Undersea Research Group, Rockville, Maryland 20852

[4]Arco Exploration Technology, 2300 West Plano Parkway, Plano, Texas 75075

Abstract

Two sand ridges on the New Jersey continental shelf were cored to determine their lithologic characteristics and possible modes of deposition. The ridge at location 1A is a nearshore, shoreface-connected sand ridge within 3 mi (4.8 km) of shore and in less than 65 ft (20.3 m) of water. The ridge at location 2 is 25 mi (40 km) from shore and in 80-115 ft (25-36 m) of water. Three main lithologic units are present in the cores from both ridges. The cores presented in this manuscript have sedimentary characteristics representative of these lithologic units and display the key differences between ridges observed in cores. The three main lithologic units are (from bottom to top): (1) *"nonfossiliferous"* sand and mud, (2) *shell-rich poorly sorted sand and mud*, and (3) *upper ridge sand*. The *"nonfossiliferous"* unit contains no macrofauna, but has traces of microfauna, massive-appearing sand layers, laminated muds, and some pebbly sand layers. The *shell-rich* unit contains numerous shell fragments and is predominantly bioturbated. Carbon-14 age determinations from the *shell-rich* unit in the nearshore ridge range from 6130 ± 120 years BP to 6360 ± 90 BP; those from the midshelf ridge range from 12,480 ± 155 years BP to 13,240 ± 180 BP. The *upper ridge sand* unit consists of stacked beds ranging in thickness from 1 to 28 in. (2.5-70 cm). Within the *upper ridge sand* unit, most beds in the nearshore ridge are up to 9 in. (22.5 cm) thick, whereas most beds in the midshelf ridge are slightly thicker, 12 in. (30 cm) or less. Within the *upper ridge sand* unit, both ridges have alternating laminated and nonlaminated (bioturbated) layers, contain fine- to medium-grained sand, and generally coarsen upward. The nearshore ridge has a slightly coarser range of mean grain size (150-400 μ) than that of the midshelf ridge (140 to 360 μ). The youngest carbon-14 age determinations from the *upper ridge sand* unit are 1480 ± 170 years BP from the nearshore ridge and 1155 ± 85 BP from the midshelf ridge.

Results from this study support the hypothesis that both the nearshore and midshelf ridges are being actively modified and possibly formed at present sea level. These conclusions support past theories on the origins of the nearshore ridges (Duane et al., 1972), but indicate that the midshelf ridge has a much more "dynamic" posttransgressional depositional history than previously surmised.

Introduction

In 1979 a consortium of oil companies headed by Cities Service initiated a geological investigation of two linear sand ridges on the New Jersey continental shelf (Figs. 1, 2) utilizing vibracore, bathymetric profile, and high-resolution seismic data. This paper focuses on the vibracore data and presents the major sedimentologic results of that study. Also included is a discussion of some of the previous work conducted on the eastern continental shelf of the United States. Ridge 1A is within 3 mi (4.8 km) of the shoreline, in less than 65 ft (20 m) water depth, and in the area of shoreline-attached sand ridges oriented approximately 30° to the shoreline (Duane et al., 1972). Ridge 2 is approximately 25 mi (40 km) offshore, in 80–115 ft (25–35 m) water depth, and has been alternately hypothesized to be a degraded barrier island (Stubblefield et al., 1984) or shelf sand ridge forming part of a shoal-retreat massif (Swift, 1973; Swift and Sears, 1974). A more recent study indicates that ridge information occurred mainly in a middle-shelf setting.

This paper describes general sedimentary characteristics of three key cores out of a total of seven coring sites at ridge 2 (middle shelf) and five coring sites from ridge 1A (nearshore). The lithofacies of these three cores are tied to the general stratigraphic framework of the two ridges using cross sections developed from all 12 cores.

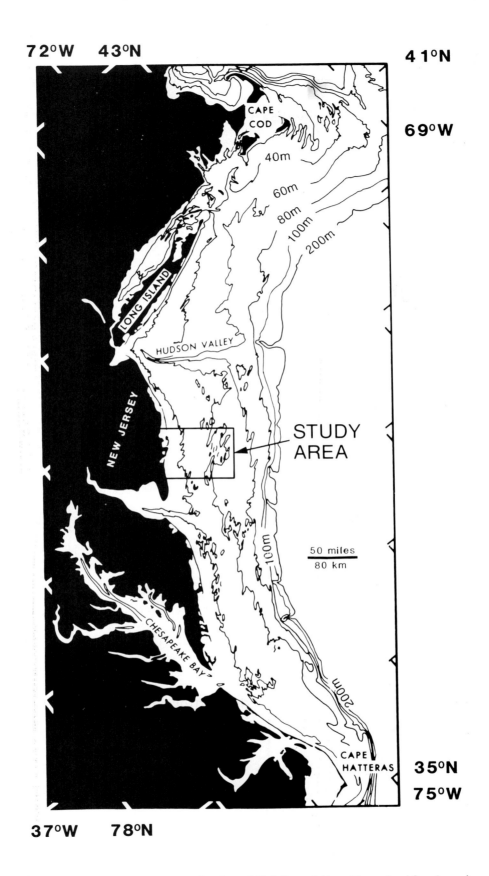

Figure 1. Bathymetric map of the Middle Atlantic shelf showing the location of the study area.

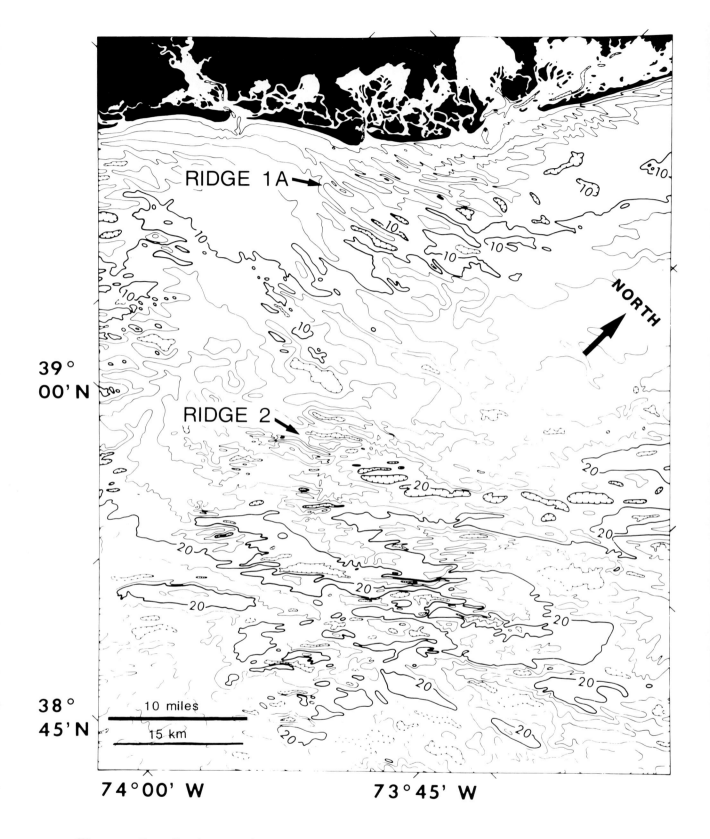

Figure 2. Bathymetric map showing locations of the nearshore, shoreface-attached ridge (ridge 1A) and the midshelf ridge (ridge 2) on the New Jersey continental shelf. Contours in meters.

Methods

Coring Procedures

Cores were obtained using an Alpine Vibracore, which basically consists of a 4-in. (10 cm), inner diameter steel pipe and a plastic core tube liner driven into the sediment with a large pneumatic hammering device at the top of the core pipe assembly. The sample was retained with a core catcher. When the rate of core penetration was judged insufficient, the vibracore was recovered and the core removed. A second vibracore was then initiated that used a water-jetting system to "jet" the core tube past the previously "resistant" layer. Because there was no meter available to determine depth of penetration, all depths below sediment surface (seafloor) of jetted cores are speculative, and only depths from the tops of these cores are given. The absence of a penetration meter also caused excessive deformation of sedimentary structures in some cored intervals owing to excessive periods of vibration after the corer reached a resistant layer or point of maximum penetration.

Laboratory Procedures

Whole cores were cut into 2-1/2 ft (0.75 m) sections and sliced longitudinally into three segments. A slab 3/4 in. (2.0 cm) thick from the center of each core section was X-ray radiographed and photographed immediately after opening. Epoxy peels were also made from these slabs. From the remaining core-slab sections, samples were taken for grain-size analysis, C-14 age dating, and porosity and permeability measurements.

Cores were described from epoxy peels, X-ray radiographs, and core slabs. Sediment coloration of the core slabs was noted immediately after opening each core. Sedimentary structures were sketched from epoxy peels and X-ray radiographs. Textural character was determined using a grain-size comparison chart, and sieve and settling tube analyses.

Sketches of the cores were made to depict how the cores actually look in epoxy peels and X-ray radiographs. Physical and biogenic structures were noted when observed. Deformation caused by coring procedures has not been removed from the core descriptions because it was difficult to distinguish many forms of coring-induced deformation from depositional sedimentary structures.

As would be expected in any unlithified, predominantly sandy sediment, some degree of deformation is present in all cores. The most common deformation feature resulting from coring is bowing of layers with vertical amplitudes, which in some cases, reaches 5 in. (13.0 cm). Other types of sediment deformation include vertical flow features and microfaulting. The top several inches of most cores was disturbed by water sloshing in the core barrel when cores were brought aboard the ship.

When the water-jetting technique was utilized to increase core penetration depth, sediment at the top of the core consistently contained a residue of coarse sand and shell material from the jetting process ("jet-lag"). The contact between the jetting artifact and the relatively undisturbed sediment below is easy to pick and is identified for all the jetted cores.

Geologic Setting

Sediments

Early studies of Atlantic shelf sediments were concerned with the source of sediment and its time of deposition. Shepherd and Cohee (1936) found little correlation between water depth and grain size, and concluded (on the basis of 200 samples) that sediment distribution on the mid-Atlantic shelf of North America reflects Pleistocene depositional patterns. Stetson (1938), in his more detailed study based on 10 transects from Nova Scotia to Florida, concluded that distribution of sediments on the shelf is adjusted to present sea level. Beginning in the 1960s, much more extensive surveys were undertaken by a Woods Hole research team. In these studies, sediment was classified as modern or relict (from earlier depositional environments) on the basis of Emery's (1952) classification. Emery concluded that the Atlantic continental shelf is covered primarily by relict sands with only a thin band (< 10 km wide) of modern detrital sediment located in the nearshore (Emery, 1965).

More recent studies have shown that the nearshore band, characterized by its fine grain size, is a fair-weather mantle of rip-current fallout forming a veneer on the shoreface (Swift et al., 1972). This veneer is stripped off during major storms, and the underlying "relict" coastal plain sands are eroded. At such times, sediments from both the fine surficial deposit and the underlying "relict" sands are swept seaward to nourish the leading edge of a transgressive shelf sand-sheet (Swift et al., 1972).

The advent of high-resolution seismic profiling on the continental shelf added a third dimension to the study of Atlantic shelf sediments.

On the basis of seismic surveys in the Baltimore Canyon area, Knebel and Spiker (1977) proposed a twofold shelf stratigraphic scenario in which a surficial sand sheet less than 11,000 years old overlies muddy-sand strata older than 24,000 years BP. This surficial sand sheet has an average thickness of 16-24 ft (5-7 m) and ranges from less than 3 to 65 ft (1-20 m) thick. Knebel and Spiker (1977) conclude that changes in thickness of the surficial shelf sand sheet are "closely related to the bottom morphology and thus to its formation."

Morphology

Morphology of the Atlantic shelf of the eastern United States is complex (Fig. 2) and is best explained in the context of the relative scales of the features present. Swift et al. (1972) categorize morphological elements of the central and southern Atlantic shelf into small- and large-scale features (Table 1). Examples of small-scale shelf features are ripples and sand waves (bedforms). The large-scale features are divided into three orders. This study examined cores from "first order" shoreface-connected (ridge 1A) and isolated (ridge 2) sand ridges that are both associated with "second order" ridge fields.

The nearshore ridges average 14 ft (4.2 m) high, 1.7 mi (2.7 km) wide, and 2.1 mi (3.4 km) in wavelength (Stubblefield et al., 1984). Some of the nearshore ridges, such as ridge 1A, are symmetrical to slightly asymmetrical, and their seaward sides are the steepest. Surficial sands on the nearshore ridges grade from coarse grained on the shoreward flank to fine grained on the seaward flank (Stubblefield et al., 1984). The midshelf ridges, such as ridge 2, are more asymmetrical than the nearshore ridges, and their seaward sides are steepest. Midshelf ridge heights and widths are slightly greater than

Table 1. Morphological elements of the central and southern Atlantic shelf (from Swift et al., 1972).

Small-scale elements:

   Ripples and sand waves

Large-scale elements:

   First Order:

      Shoreface-connected ridges and swales (i.e., ridge 1A)

      Isolated ridges and swales (i.e., ridge 2)

   Second Order:

      Cape-associated shoals

      Inlet-associated shoals

      Ridge fields

   Third Order:

      Shoal-retreat massifs

      Shelf-transverse valleys

      Cuestas

      Deltas

      Scarps

---

those of the nearshore ridges. Surficial grain-size distribution of the sand fraction is symmetrical, and the coarsest sand is generally near the upper shoreward flank (Stubblefield et al., 1984). Wavelengths of the midshelf ridges are less and average 1.4 mi (2.3 km). Orientation of the ridges is different; the nearshore ridges average 25° with respect to the coastline, and the midshelf ridges are approximately coast-parallel (Stubblefield et al., 1984).

Shelf Processes

The continental shelf of New Jersey is a storm-dominated shelf (Duane et al., 1972). Regarding shoreface-connected ridges such as ridge 1A, Duane et al. (1972) state "storm current and wave trains associated with them are dominant forces shaping the shoals." The most significant storms on the Atlantic shelf are extratropical (winter) storms. According to Harrison et al. (1967), successive midlatitude lows passing along the Atlantic seaboard during the winter months generate strong northeast winds. The resulting wind stress, combined with falling temperatures, destroys water stratification, allowing wind-driven currents to affect the bottom. Harrison et al. (1967) also noted that a net southerly drift is established owing to the onshore component of the winter storm winds causing set-up of water along the coastline. Using current-meter surveys of three sites on the outer New Jersey shelf, Butman et al. (1979) found winter energy levels much higher than summer levels. Measurements taken from December 1975 to February 1976 showed along-shelf currents to average 15 cm/s; maximums of 40 cm/s occurred once a month. Measurements taken from June to July showed average currents of less than 8 cm/s. The only high-velocity current events recorded during summer months were brief

10

(3-25 h) and attributed to internal wave packets generated by semidiurnal tides at the shelf edge.

General Lithologic Units

Three major lithologic units are present in both the nearshore (ridge 1A) and midshelf (ridge 2) sand ridges. In this study they have been designated *"nonfossiliferous" mud and sand, shell-rich poorly sorted sand and mud*, and *upper ridge sand*. Although these units are present in both ridge systems, at least the upper two units are not correlatable between ridges. The only relationship these units have between ridges is that they record the presence of similar depositional processes. All of these units are present within core V-13 (Figs. 3, 18). Major sedimentary characteristics of these units are described below on the basis of all cores taken during the 1979 cruise. Figure 4 shows locations of the cores on the nearshore ridges and Figure 6 on the midshelf ridges. Figures 5 and 7 show composite cross sections through the nearshore and midshelf ridges.

"Nonfossiliferous" Mud and Sand Unit

Nearshore ridge. Within the nearshore ridge, the *"nonfossiliferous" mud and sand* unit is present at the base of cores V-13 (Figs. 3, 18), V-12A, and V-14A (Fig. 5). Macrofauna are absent from this unit, but microfauna are present (S. Culver, pers. comm.). Sand layers consist primarily of fine- to medium-grained sand (220-550 μ mean grain-size range), but some granule- to pebble-size grains are also present. These sand layers are massive-appearing or subhorizontally laminated, as in V-13 (Figs 3, 18). The sand layers are interlayered with beds of mud and muddy sands. The mud layers

11

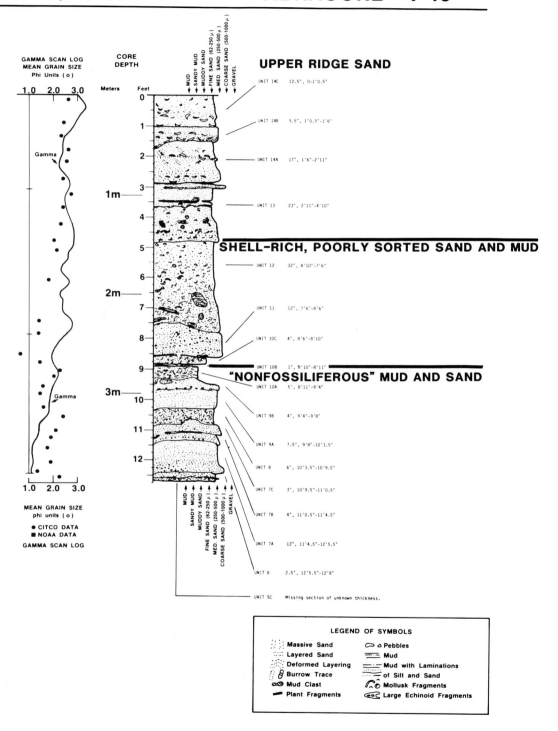

Figure 3. Description of core V-13 showing sedimentary structures, grain-size trends, and core gamma.

VIBRACORE V-13

UNIT 14C   12.5", 0-1'0.5"

Fine-grained sand that fines upward (200-150µ) and has shell fragments (5%) scattered throughout but most abundant near base. There are no structures other than a diffuse, swirly pattern in the x-ray radiograph. This unit appears denser than Unit 14B in the radiograph. The unit is yellowish gray (5Y7/2) below 0'11" and above 0'2" and is olive gray (5Y4/1) to medium dark gray (N4) between.

UNIT 14B   5.5", 1'0.5"-1'6"

Fine-grained sand (250µ) with shells (5-10%) throughout but more abundant at the unit base. The lower contact is very irregular with a relief of 4". Shells above the base are arranged in parallel layers inclined 20° to 30°. This is the only apparent structure in this olive gray (5Y4/1) to dark gray (N3) and yellowish gray (5Y7/2) above 1'4" interval.

UNIT 14A   17", 1'6"-2'11"

Medium- to fine-grained sand that fines upward (300µ to 180µ) and contains shells near base and in a pocket at 2'1". This pocket of shells is oriented diagonally (approximately 45°), is 1" at its widest point and is probably an infilled burrow. Along with is 1" wide tract there are 1/8" to 1/4" wide (burrow) traces present in x-ray radiographs between 1'6" and 2'4". The sand is olive gray (5Y4/1) to dark gray (N3).

UNIT 13   23", 2'11"-4'10"

Fine-grained sand that fine upward (250µ to 180µ) with numerous shells mostly below 4'0" and concentrated below 4'5". Some bivalve shells are articulated. Two, near-horizontal, mud layers are present at 3'8" an 3'6". These 1/2" thick layers may be burrows. Thin (1/8" wide) vertical to diagonally oriented, linear mud pockets at 4'4" to 4'1" are probably mud lined burrows. Diagonal, 1/8" wide burrow traces are visible in x-ray radiographs between 4' an 3'6". Also a 1" wide, near horizontal, discontinuous pocket of coarse sand and shells is at 3'0" and this is probably a burrow fill. No physical structures are observed in this olive gray (5Y4/1) sand with darker, mud pockets and layers.

UNIT 12   32", 4'10"-7'6"

Mud rich, medium-grained sand that fines upward (average size of sand component varies from 350µ to 250µ), contain shell debris (5%) and mud pockets (clasts) between 6'0" and 7'6". Articulated bivalve at 7'3" appears to be in life position. No structures are visible within unit, probably from bioturbation. Interval is yellowish gray (5Y7/2) below 7'3" and olive gray (5Y4/1) to olive black (5Y2/1) above.

UNIT 11   12", 7'6"-8'6"

Medium-grained sand (350µ) with granules (below 8'0") and one shell fragment at base. Interval is well layered (crossbedded?) below 8'0" with layering less apparent above. Sand is olive gray (5Y4/1).

UNIT 10C   4", 8'6"-8'10"

Structureless, sandy (200µ) mud with very few granules (<1%) and topped by laminated layer (1/2" thick) of mud and sand. There are a few small carbonaceous fragments at base of this olive gray (5Y4/1) unit.

UNIT 10B   1", 8'10"-8'11"

Laminated, olive gray (5Y4/1) silty clay layer.

UNIT 10A   5', 8'11"-9'4"

Sandy (200µ) mud that is olive gray (5Y4/1) in color and appears to have disrupted laminae (in x-ray radiographs). Less dense pockets (1/8") in x-ray radiographs may be burrows. Mud clast is at 9'1".

UNIT 9B   4", 9'4"-9'8"

Medium-grained sand (350µ) with pebbles (<1cm) and granules at base. Layering is not apparent but unit was disrupted when slabbed. Top of unit is very irregular with a pocket of muddy sand at 9'5". Interval is olive gray (5Y4/1) mottles.

UNIT 9A   7.5", 9'8"-10'3.5"

Fine-grained sand (220-250µ) that is layered throughout. Layers appear diffuse in x-ray radiographs and are bowed. Sand is yellowish gray (5Y7/2).

UNIT 8   6", 10'3.5"-10'9.5"

Sandy (350µ) mud with some granules (1-2%) scattered throughout the structureless interval. Mud is slightly darker than olive gray (5Y4/1).

UNIT 7C   3", 10'9.5"-11'0.5"

Fining upward, medium-grained sand (350-250µ) that is deformed throughout. Upper contact is bowed over 3". Sand is yellowish gray (5Y7/2).

UNIT 7B   4", 11'0.5"-11'4.5"

Fining upward, medium-grained sand (300µ) with vague traces of layering that is bowed. Some granules are present in the yellowish gray (5Y3/2) sand. Unit is topped by a thin lamination of olive gray (5Y4/1) mud.

UNIT 7A   12", 11'4.5"-12'5.5"

Medium-grained sand that fines upward (400µ to 250µ) and is layered below 11'9". Granule-size grains are below 12'1". Layering is bowed. Sand is medium gray (N5) below 12'8" and yellowish gray (5Y7/2) with olive gray (5Y4/1) laminations.

UNIT 6   2.5", 12'5.5"-12'8"

Muddy, fine-grained sand (250µ) that is layered. Pebble (1cm) is in slightly coarser lamina at 12'7". Sediment is olive gray (5Y3/1) to olive black (5Y2/1).

UNIT 5C   Missing section of unknown thickness.

13

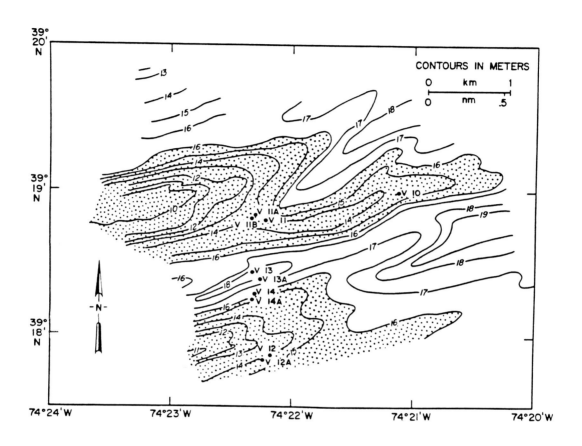

Figure 4. Bathymetry of the nearshore ridge (location 1A) and locations of vibracore stations. Contour interval is in meters. Landward is toward the upper left-hand corner of the diagram.

14

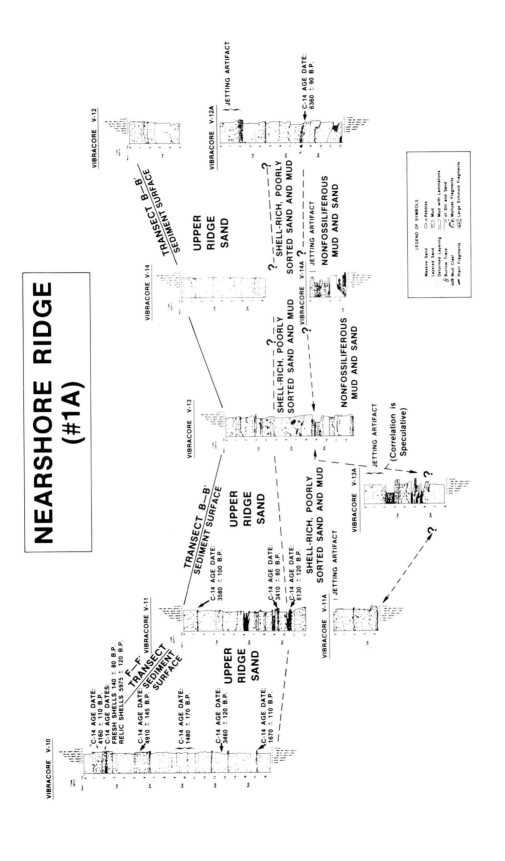

Figure 5. Composite cross section of all cores from the nearshore ridge (1A) showing lithologic correlations between cores and lithostratigraphic positions of C-14 age data. See Figure 4 for location of cores. Cores are only approximately arranged with regard to water depth.

15

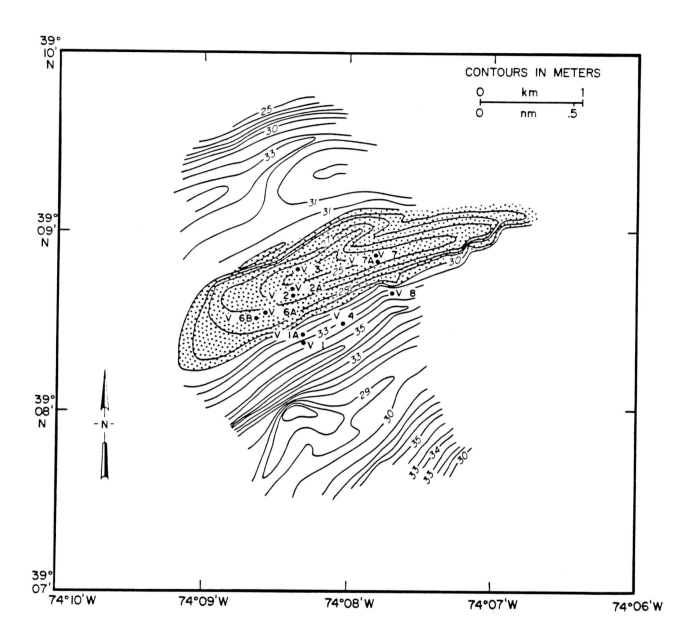

Figure 6. Bathymetry of midshelf ridge (location 2) and locations of vibracore stations. Contour interval is in meters. Landward is toward the upper left-hand corner of the diagram.

16

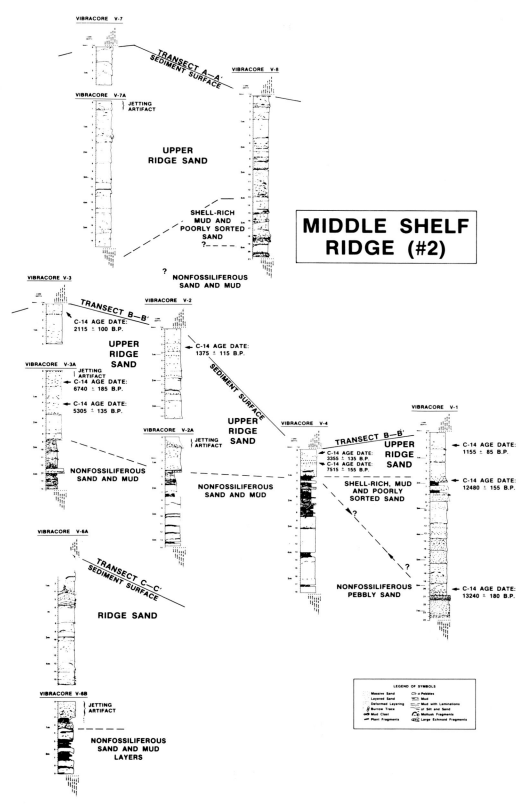

Figure 7. Composite cross section of all cores from the midshelf ridge (2) showing lithologic correlations between cores and lithostratigraphic positions of C-14 data. See Figure 6 for location of cores. Cores are only approximately arranged with regard to water depth.

are laminated with only rare burrow traces. The muddy sand layers appear "churned" and may be bioturbated (over 75% burrowed by volume) or disturbed by coring.

Midshelf ridge. Within the midshelf ridge, the *"nonfossiliferous" mud and sand* unit is present at the base of cores V-2A (Figs. 9, 16), V-3A, V-4, V-6B, and V-8 (Fig. 7). The unit consists of interlayered sand and mud with no observable macrofaunal remains and only rare burrow traces. A preliminary examination for microfauna revealed that only one sample out of the three examined contained foraminifera. The foraminifera assemblage present indicates a bay or marsh environment (J. Etter, pers. comm.).

Shell-Rich Poorly Sorted Sand and Mud Unit

Nearshore ridge. Within the nearshore ridge the *shell-rich poorly sorted sand and mud* unit is present in cores V-13 (Figs. 3, 18), V-11, V-11A, V-12A, and V-13A (Fig. 4). Sand beds within this unit contain fine- to coarse-grained sands (mean grain-size range of 160-800 μ with shell fragments and granules and/or pebbles. Beds generally fine upward and are massive-appearing, except for some apparent cross-bedding in V-13 (Figs. 3, 18; 8.0-8.5 ft; 2.5-2.6 m) and subhorizontal laminations in V-13A (Fig. 5). The massive appearance of most of the sand in this unit is probably due to bioturbation. Recognizable burrow traces are present. Mud and sandy mud layers consist of laminated silts and clays, and bioturbated sands and muds, as in V-13 (Figs. 3, 18; 8.9-9.3 ft; 2.8-2.9 m).

C-14 dates of bulk shell samples from this unit within the nearshore ridge are 6130 ± 120 years BP (V-11; 10.5-10.7 ft; 3.2-3.3 m)

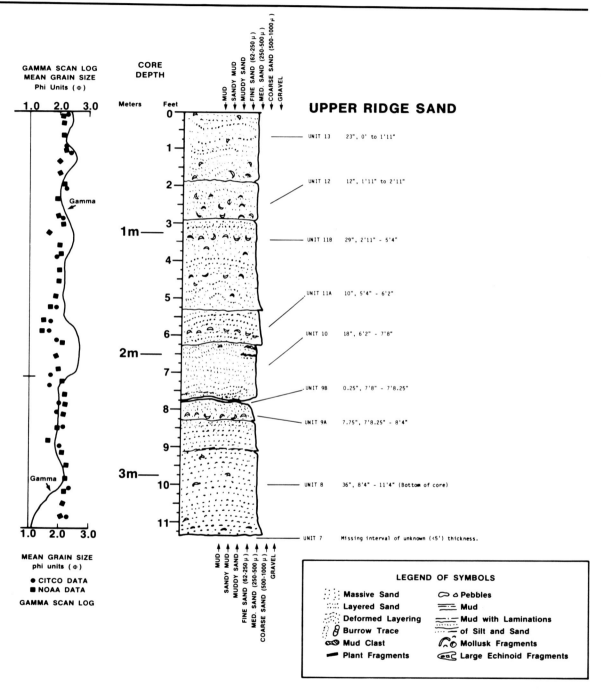

Figure 8. Description of core V-2, showing sedimentary structures, grain-size trends, and core gamma.

VIBRACORE V-2

UNIT 13   23", 0' to 1'11"

Fine-grained sand 225μ) with skeletal fragments concentrated at 1'8" and 0'2" as well as thinly dispersed throughout.   The interval is vaquely laminated (in x-ray radiographs in over 30% of the unit.   The sand is yellowish gray (5Y7/2) to moderate yellowish brown (10YR5/4) with dusky yellowish brown (10YR2/2) mottling.

UNIT 12   12", 1'11" to 2'11"

Vaguely laminated (in x-ray radiographs) fine-grained sand (225μ) with skeletal fragments (echinoid and mollusk) more abundant below 2'3". The sand is light olive gray to yellowish gray (5Y7/2).

C-14 dating of skeletal fragments from this interval indicate an origin of 1375 $\pm$ 115 B.P.

UNIT 11B   29", 2'11" - 5'4"

Medium-grained sand (275-250μ) fining up to fine grained sand (225μ). The unit is vaguely layered in 30% of the interval. Skeletal fragments (echinoid and mol-lusk) are concentrated in layers at 4'5" and 3'4". The sand is light olive gray (5Y5/2) or light olive gray to yellowish gray (5Y7/2) with olive gray (5Y4/1) or dark gray (N3) mottling.

UNIT 11A   10", 5'4" - 6'2"

Well laminated, medium-grained sand that is slightly fining upward (300μ to 250μ). Mollusk fragments are concentrated in a lamination at 5'9". The sand is olive gray (5Y4/1) below 6'0" and yellowish gray (5Y47/2) above.

UNIT 10   18", 6'2" - 7'8"

Fine-grained sand (225μ) that is slightly coarser at base (250μ). Unit is vaguely layered at top (6'2" to 6'7") and base (7'4" to 7'8"). Shell fragments are scat-tered throughout but more concentrated in layers at 6'5" and 6'7". Some plant fragments are present. The sand is yellowish gray (5Y7/2) with olive gray (5Y4/1, 5Y2/1, 10Y4/2 or 5Y6/1) mottles between 7'8" and 6'4" and olive gray (5Y4/1) between 6'2" and 6'4".

UNIT 9B   0.25", 7'8" - 7'8.25"

Olive gray (5Y4/1) mud with small shell fragments and fibrous and rod shaped plant fragments.

UNIT 9A   7.75", 7'8.25" - 8'4"

A fining upward fine-grained sand (225μ to 180μ) with numerous skeletal fragments throughout. Mollusk and large (+1cm) echinoid fragments are present near the base (8'1" to 8'4"). Smaller shell fragments and numerous small (<1cm long) echinoid spines are present near the unit top, dispersed in the sand and also concentrated in a pocket at 7'9". A few plant fragments are present within the unit. The sand is light olive gray (5Y6/1) to olive gray (5Y4/1).

UNIT 8   36", 8'4" - 11'4" (Bottom of core)

Well laminated, fine-grained sand (250μ), that is slightly coarser at the base of the core (bimodal, 250 and 500μ) and finer at the unit top (225μ). The laminations which are very apparent in the x-ray radiographs are fuzzy (bioturbation?) and bowed (coring artifact). The sand is medium dark gray (N3.5) or light olive gray (5Y6/1) with olive gray (5Y4/1) and medium dark gray (N4) mottling.

UNIT 7   Missing interval of unknown (<5') thickness.

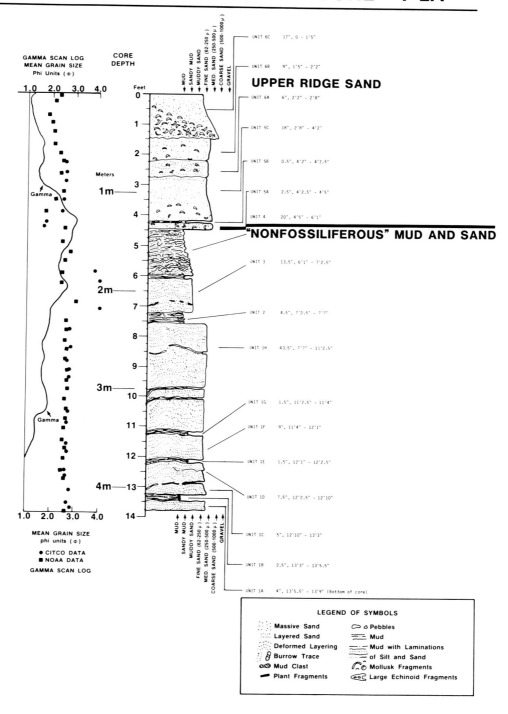

Figure 9. Description of core V-2A, showing sedimentary structures, grain-size trends, and core gamma.

VIBRACORE V-2A

UNIT 6C   17", 0 - 1'5"

Interval consists of sediment disturbed by jetting. Shell lag is at base and is covered by fine-grained sand that is yellowish gray (5Y2/2) in color.

UNIT 6B   9", 1'5" - 2'2"

Fine-grained sand (180μ) with a few skeletal fragments (echinoid and mollusk). Unit appears mottled in x-ray radiographs and also disturbed by coring. Sediment is olive gray (5Y4/1) to light olive gray (5Y6/1).

UNIT 6A   6", 2'2" - 2'8"

Fine-grained sand (200μ) with a few mollusk fragments (pelecypod and gastropod). Unit has mottled appearance in x-ray radiographs which may be from bioturbation. Sediment is olive gray (5Y4/1) to medium dark gray (N4).

UNIT 5C   18", 2'8" - 4'2"

Fining upward medium-grained sand (bimodal with 200 and 450μ) to fine grained sand (180μ) with some shell fragments near the base (3'6" to 4'2"). Traces of layering present near base (below 3'4"). Sediment has mottled appearance in x-ray radiographs above 3'4". Sediment color is olive gray (5Y4/1) to medium dark gray (N4).

UNIT 5B   0.5", 4'2" - 4'2.5"

Mixed sand, silt and mud lamination topped by a thin lamination of clay. Color is yellowish gray (5Y7/2). Unit boundaries are not bowed like all other unit boundaries Unit was probably an artifact of core handling.

UNIT 5A   2.5", 4'2.5" - 4'5"

Coarse-grained sand (500μ) with a few mollusk fragments (1-3%). The sand is structureless and is yellowish gray (5Y7/2) is color.

UNIT 4   20", 4'5" - 6'1"

Interlayered laminations and very thin beds of silty clay with laminations of sand (90μ). Sand layer or pocket at 5'9" consists of 110μ sand. Over 40% of vertical interval contains vertical burrows. Most of these sand filled or sand outlined burrows are less than 1/16" in diameter but some range up to 1/2" in diameter (5'3" core depth). Sand content (thickness and abundance of laminae) decreases upward within the unit. The muds are medium dark gray and the sands are dark gray (N3).

UNIT 3   13.5", 6'1" - 7'2.5"

Muddy sand whose sand component increases in grain size upward (from an average of 70μ to 180μ). About 80% of unit contains very disturbed laminations of sand and silty clay (bioturbation of coring artifact?). An interval of laminated muddy sand is present near 6'7" to 6'10". These laminae are disturbed by vertical burrows (1/4" to 1/8" in diameter). The top 1/2" of this unit contains thin (<1/16" diameter) vertical burrows and appears in x-ray radiographs to be bioturbated. The sediment is medium gray in color.

UNIT 2   4.5", 7'2.5" - 7'7"

Silty clay with laminations of muddy, fine-grained sand and vertical sand fill burrows in the top 1" of the layer. The vertical burrows are short (1/2" to 1" long) and either almost 1/4" diameter or less than 1/16" in diameter. The mud is olive gray (5Y4/1) to medium dark gray (N4) and the sand is olive gray (5Y4/1).

UNIT 1H   43.5", 7'7" - 11'2.5"

Laminated fine sand (180μ) with laminations of silty clay. Both the sand and mud laminations are very deformed from coring. Brownish gray (5YR4/1) mud clasts are present at 11'0.5". The sand is moderate yellow brown; (10YR5/4) with dark yellow brown (10YR2/2) mottles (10'8" to 11.205); medium gray (N5) to olive gray (5Y4/1) with medium dark gray layering (N4) (10' to 10'8"); moderate yellow brown with dark yellow brown (10YR4/2) mottles (8'10" to 10'); an olive gray (5Y4/1) with dark gray (N3) layering (7'7" to 8'10"). The mud layers are brownish gray (5YR4/1) to brownish black (5YR2/1) (10' and 9'8") and yellowish gray (5Y7/2).

UNIT 1G   1.5", 11'2.5" - 11'4"

Silty clay with sand laminations. Lenticular sand fill may be an injection resulting from coring. Mud is medium dark gray (N4).

UNIT 1F   9", 11'4" - 12'1"

Fine-grained sand (180μ) with numerous laminations that are highly deformed by coring. Unit is mottled with dark gray (N3) medium dark gray (N4), moderate yellow brown (10YR5/4) medium gray (N5) and olive gray (5Y4/1).

UNIT 1E   1.5", 12'1" - 12'2.5"

Silty clay laminated with sand. Laminations are irregular in thickness. Thin (<1/8") vertical "cracks" that are infilled with sand are also present. The upper boundary of the mud is irregular while the lower boundary is bowed (by coring) but regular. Sediment is medium dark gray (N4) to olive gray (5Y4/1).

UNIT 1D   7.5", 12'2.5" - 12'10"

Fining upward fine-grained sand (200μ to 180μ) with numerous (parallel?) laminations which are deformed from coring. Sediment is medium dark gray to dark gray (N3.5) with dark gray (N3) mottles, except for the top 2" which are moderate yellow brown (10YR5/4) with dark gray (N3) mottling.

UNIT 1C   5", 12'10" - 13'3"

Fining upward fine-grained sand (200μ to 180μ) that is laminated (very deformed by coring). Unit 1C and 1D are separated by a thin lamination of silty clay. Sand is olive gray (5Y4/1) to dark gray (N3) and the mud is medium gray (N5).

UNIT 1B   2.5", 13'3" - 13'5.5"

Silty clay laminated with sand. The laminations are lenticular. Sediment is medium dark gray (N3.8)

UNIT 1A   4", 13'5.5" - 13'9" (Bottom of core)

Fine-grained (200μ) sand with trace of laminations. Interval becomes more muddy upward and includes laminations of clay. Sediment is dark yellow brown (10YR4/2) to pale yellow brown (10YR6/2).

and 6360 ± 90 BP (V-12A; 7.8-8.2 ft; 2.4-2.5 m). The accuracy of these dates is doubtful since portions of the shells used for dating were probably reworked. Most sea level curves place the nearshore region above sea level from 6100 to 6500 years BP.

Midshelf ridge. Within the midshelf ridge, the *shell-rich poorly sorted sand and mud* unit is present only in cores V-1 and V-8 (Fig. 7). Within this unit, mud layers are interbedded with shell-rich, poorly sorted sands. Within core V-1 (5.7-20.7 ft; 1.7-6.3 m) the *shell-rich* unit sharply overlies a nonfossiliferous pebbly sand bed (Fig. 7). C-14 dates from shell materials in V-1 indicate dates of 13,240 ± 180 years BP (V-1, 19.2-20.7 ft; 5.8-6.3 m) and 12,480 ± 155 BP (V-1, 5.7-6.5 ft; 1.7-2.0 m). The samples used for dating probably included some reworked shells since most sea level curves (i.e., Moore and Curray, 1974) place the midshelf region above sea level during the 12,300 to 13,400 BP period.

Upper Ridge Sand Unit

Nearshore ridge. The *upper ridge sand* unit is a generally coarsening-upward sequence of fine- to medium-grained sand (150-400 µ) having numerous stacked beds averaging 9 in. (20 cm) thick. Most cores of this unit coarsen upward, such as V-10 (Fig. 10). Only core V-13, located on the flank of a ridge, does not coarsen upward (Fig. 3). Beds within the *upper ridge sand* unit range in thickness from 2 to 18 in. (5-45 cm). Bed boundaries are delineated on the basis of changes in texture or character of sedimentary structures. A common change in sedimentary structures along a bed boundary is one of

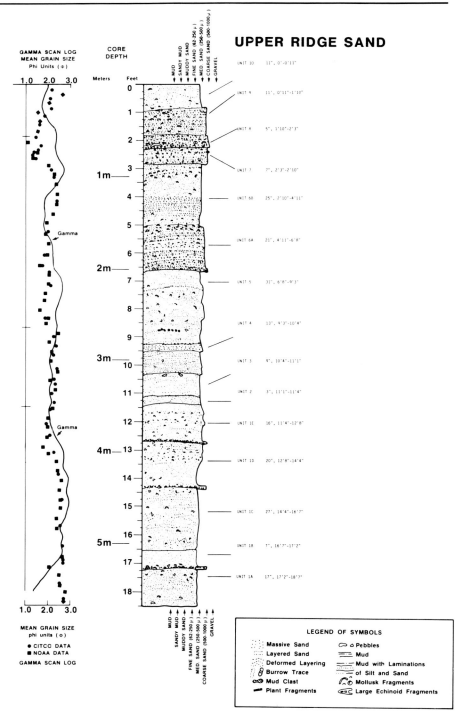

Figure 10. Description of core V-10, showing sedimentary structures, grain-size trends, and core gamma.

VIBRACORE V-10

UNIT 10   11", 0'-0'11"

Sand, fine-grained (200μ) with only a trace of shells. Massive appearing in upper 2/3 and subhorizontal laminated in lower 1/3. Probably part of fining up sequence which includes Units 8, 9, 10. Yellow gray (5Y7/2) to light olive gray (5Y5/2).

UNIT 9   11', 0'11"-1'10"

Medium-grained sand (300μ) with abundant shells (25%) that are slightly more concave up. Slightly deformed subhorizontal to very low angle laminated. Laminasets are 1/2" thick. Sand is yellowish gray (5Y7/2) with light brown (5YR4/2) bands, except for lower part which is medium dark gray.
C-14 date from shell material between 0'11" to 1'10" is 4160 + 110 B.P.

UNIT 8   5", 1'10"-2'3"

Medium-grained sand (350μ) that is very shelly (60%) with shells poorly sorted and up to 3" long. No bedding visible. Unit is probably a storm deposit. Sand is yellowish gray (5Y8/1). Unit is coarsest in core.
C-14 date from fresh shell material between 1'11" to 2'3" is 140 + 90 B.P.
C-14 date from relict shell material between 1'11" to 2'3" is 5975 + 120 B.P.

UNIT 7   7", 2'3"-2'10"

Sand, medium-grained (400μ) with high angle (30°) planar cross bedded 3/8" laminasets and 10-15% shells. Strong grain size contrast between laminasets. Medium dark gray (N4) with yellow-gray (5Y8/1) banding.

UNIT 6B   25", 2'10"-4'11"

Sand, fine- to medium-grained (250μ), similar to unit below except less shelly and less oxidized near the top. Zone of probable worm burrows at 3'5"-3'9" (1/16" dia.) Diffuse to well laminated in about 60% of unit; visible on x-ray radiographs and peel. On peel massive appearing upper portion. Sharp upper contact. Upper 8" dark gray (N3) to olive gray (5Y4/1), below yellowish gray (5Y8/1) with dark gray (5Y4/1) laminations.

UNIT 6A   21", 4'11"-6'8"

Shelly sand (350μ), 15% shells, three zones of up to 40% shells. Shells about equally concave up and down orientation. Diffusely laminated on x-ray. Deformation, bowed up to 1 1/2". 1/4"-1" laminasets visable on peel. Yellowish gray (5Y8/1) with olive gray (5Y4/1) laminations.

C-14 date from relict shell material between 4'11" to 5'11" is 4910 + 145 B.P.

UNIT 5   31", 6'8"-9'3"

Medium-grained sand (300μ) that is massive appearing on peel and slightly diffuse in upper 3" (burrowed?) on x-ray radiographs. Bowing (2.5") of lamina near middle, otherwise apparently undisturbed. Trace of shells near base; top 1' almost shell free. Sharp upper contact. Upper 21" medium dark gray (N4) to olive gray (5Y4/1); lower portion yellowish gray (5Y7/2) with olive gray (5Y4/1) laminations.

UNIT 4   13", 9'3"-10'4"

Fine- to medium-grained sand (200-300μ) that is slightly coarser top 3". Conspicuously laminated on both core and x-ray radiographs. Laminasets mostly 1/8" thick and were initially subhorizontal to low angle parallel laminated. Interlaminated (180-350μ) layers. Laminae not truncated. One 1/4" diameter burrow. Only a trace of shells. Yellow gray (5Y7/2) with olive gray (5Y4/1) laminations.

UNIT 3   9", 10'4"-11'1"

Fine-grained sand (200μ) with very few shells. Not highly disturbed, but "massive appearing" on x-ray radiographs, bowed 1 1/2" at top 1" possibly burrowed. Yellow gray (5Y7/2) with olive gray (5Y4/1) laminations.
C-14 date from shell material between 8'11" to 11'3" is 1480 + 170 B.P.

UNIT 2   3", 11'1"-11'4"

Fine-grained sand (200μ) that appears to be bioturbated in x-ray radiographs. Burrow traces are 1/8" in diameter and over 1" in length. Sand is medium and dark gray (N4) to olive gray (5Y4/1).

UNIT 1E   16", 11'4"-12'8"

Fine-grained sand (250-300μ). 4 bowed (1") laminae otherwise massive appearing in both peel and x-ray, not highly disturbed. 1/2" zone of higher shell concentration at base. Similar to underlying unit. Medium dark gray (N4) to olive gray (5Y4/1) with light olive gray (5Y6/1) layers.

UNIT 1D   20", 12'8"-14'4"

Fine-grained sand (150-200μ). Base of unit 1/2" zone of shells. Upper 9" slightly lighter gray (less oxidized) and more shelly (5-8%) than below. Shells slightly more commonly concave down. Possible 3" burrowed zone at 13'2". Similar to units below. Yellowish gray to light olive gray (5Y7/1).

C-14 date from shell material between 12'8" to 14'3" is 3460 + 120 B.P.

UNIT 1C   27', 14'4"-16'7"

Fine-grained sand (150-200μ) that is mildly disturbed by coring, above 14'9" bowed (1"). Base of unit is slightly shellier. Unit is structureless and light olive gray (5Y6/1).

UNIT 1B   7", 16'7"-17'2"

Fine-grained sand (150μ) with fewer shells than unit 1A (3%). Base has shelly layer which is bowed (1.5"). Unit is structureless and medium dark gray (N4) to olive gray (5Y4/1) except for light olive gray (5Y6/1) band at base.
C-14 date from shell material between 16'11" to 17'9" is 1670 + 110 B.P.

UNIT 1A   17", 17'2"-18'7"

Fine-grained sand (150μ) that is massive appearing on peel. In X-ray radiographs the layering is bowed (1.5') in upper and lower portions and highly disturbed in the middle. Numerous broken shells (5%) with most in upper half. Sand is medium dark gray (N4) to olive gray (5Y4/1).

27

laminated sands overlying apparently massive/bioturbated sands (Figs. 10, 17).

Physical sedimentary structures in the *upper ridge sand* unit are primarily low angle or subhorizontal laminations (Figs. 10, 17). Some apparent cross-bedded intervals occur, as in V-10 (Figs. 10, 17; 2.2-2.8 ft; 0.7-0.8 m) and V-11 (4.5-5.5 ft; 1.4-1.7 m).

Almost totally bioturbated intervals and intervals containing distinct burrow traces are common within this unit. Bioturbated and massive-appearing intervals increase in abundance with increasing depth of burial (Figs. 10, 17). An exception to this is V-13, which is bioturbated throughout the *upper ridge sand* unit (Figs. 3, 18). Recognizable burrow traces consist of (1) burrows infilled with mud pellets and shells (V-11, 8.2-8.9 ft; 2.5-2.7 m; and 6.5-7.3 ft; 2.0-2.2 m); (2) a burrow filled with articulated bivalves in V-10 (Figs. 10, 17; 8.8 ft; 2.7 m); and (3) 1/8 in. (0.3 cm) diameter mud-lined burrows and 1 ft (2.5 cm) diameter traces within V-13 (Figs. 3, 18).

C-14 dates of bulk shell material from the *upper ridge sand* unit in nearshore ridge range from 4160 ± 110 years BP to 1480 ± 170 BP (V-10, Fig. 5). The bulk shell samples are an admixture of different aged shells, as revealed by C-14 dating of fresh shells (140 ± 90 BP) and abraded and worn shells (5975 ± 120 BP) from the same interval in V-10 (1.9-2.2 ft; 0.6-0.7 m).

Midshelf ridge. The *upper ridge sand* unit of the midshelf ridge is a generally coarsening-upward sequence of fine- to medium-grained sand (170-400 µ) within stacked, laminated, and nonlaminated beds. Nonlaminated beds are probably bioturbated, but core deformation inhibits positive identification of burrows. Cores V-2 (Figs. 9, 16),

28

V-3A, V-6A, V-7, and V-7A (Fig. 7) have interlayered laminated and nonlaminated beds.   V-8 is only laminated in the top 18 in. (45 cm) and is completely bioturbated below that within the *upper ridge sand* unit (Fig. 7).   Thickness of beds ranges from 2 to 28 in. (5-70 cm), and most are 12 in. (30 cm) or less.   Some of the boundaries between beds may be obscured, however, owing to bioturbation and core deformation. Laminated intervals within the cores consist primarily of subhorizontal laminations (as best as can be determined with the degree of core deformation present).   Cross-bedding is present in some cores (Fig. 7, V-8, 0.8-1.5 ft; 0.2-0.4 m).

C-14 dates of bulk shell (mixed fresh and worn shell) material from the midshelf ridge range from 7515 ± 155 years BP in V-4 (1.0-1.8 ft; 0.3-0.5 m) to 1155 ± 85 BP in V-1 (1.2-2.5 ft; 0.4-0.8 m; Fig. 7).

Discussion

Comparison of Nearshore and Midshelf Sand Ridges

The lithologic characteristics of the two sand ridges examined in this study are both similar and different.   The *upper ridge sand* units are generally similar in both ridges, but the underlying *shell-rich poorly sorted sand and mud* unit and the *"nonfossiliferous"* mud and sand unit differ.   These differences are probably due to differences in water depth at the time of deposition, or in the case of the *"nonfossiliferous"* unit, to deposition in a range of marginal marine to nearshore marine environments.   One striking similarity between ridges is the sharp contact between the *upper ridge sand* unit and the underlying unit (Figs. 5, 7).   This sharp contact denotes a break in time of some duration and an abrupt change in depositional processes.   Although the *shell-rich* and *"nonfossiliferous"* units commonly make up part of the relief-forming portion of both ridges, they appear not to be genetically related to the overlying *upper ridge sand* unit. Figure 11 graphically depicts how the units underlying the upper ridge sand unit are incorporated into the relief-forming portion of a ridge.

The *upper ridge sand* unit in both nearshore and midshelf locations consists of stacked beds ranging from 1 to 28 in. (2-70 cm) thick. Most beds in the nearshore ridge are 9 in. (20 cm) thick or less, whereas in the midshelf ridge they are generally 12 in. (30 cm) thick or less. The *upper ridge sand* unit from both ridges consists of alternating laminated and nonlaminated (bioturbated) beds, a sequence similar to that found in ancient and modern shoreface and inner-shelf sand deposits (Howard, 1972; Howard and Reineck, 1981). Laminations within the *upper ridge sand* unit are generally low angle to subhorizontal. Inclined or cross-bedded intervals are present but rare.

Upper ridge sand units from both ridges contain fine- to medium-grained sand and generally tend to coarsen upward. Sieve analysis indicates that the nearshore ridge (150-400 μ) is slightly coarser than the midshelf ridge (140-360 μ). Both ridges are capped by coarser grained beds (350-400 μ). The *upper ridge sand* unit within the midshelf ridge generally overlies the *"nonfossiliferous" mud and sand* unit, whereas the *upper ridge sand* unit in the nearshore ridge always overlies the *shell-rich poorly sorted sand and mud* unit. Within the midshelf ridge the *shell-rich* unit is present only in cores V-1 and V-8.

Comparison of C-14 dates from bulk shells within the *shell-rich* unit indicates that dates are much older within the midshelf ridge (V-1, 12,480 ± 155 to 13,240 ± 180 years BP) than in the nearshore ridge (V-11 and V-12A, 6130 ± 120 to 6360 ± 90 years BP). Youngest C-14

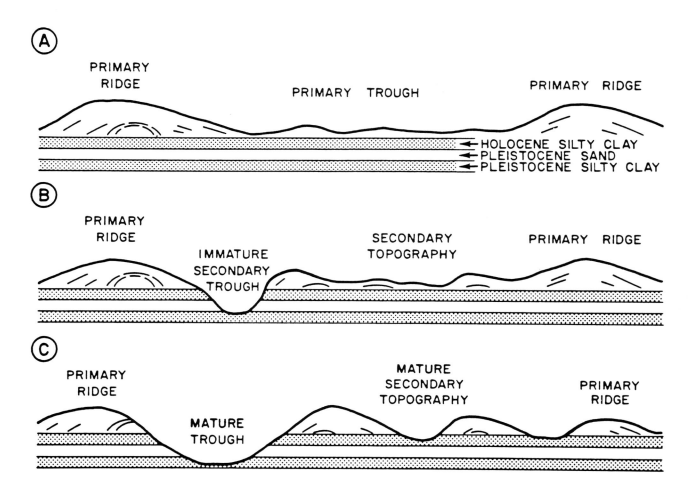

DEVELOPMENT OF RIDGE TOPOGRAPHY

Figure 11. Sequential model for midshelf ridge development by
Stubblefield and Swift (1976). [A] Large-scale ridges are
initiated and grow by vertical and lateral aggradation.
Large-scale primary troughs are zones of nondeposition.
[B] Small-scale, secondary troughs are incised into older
substrate. [C] Secondary troughs widen and aggradation
of secondary ridges is fed in part by sand released by
trough erosion.

dates from bulk shell material within the *upper ridge sand* unit are similar in both ridges. From the midshelf ridge, the youngest date is 1155 ± 85 years BP (V-1, 1.2-2.5 ft; 0.4-0.8 m; Fig. 7). From the nearshore ridge the lowest date is 1480 ± 170 BP (V-1, 8.9-11.2 ft; 2.7-3.4 m; Fig. 5).

Genesis of the Nearshore Ridge

The nearshore ridge lies within the zone of actively forming shoreface-connected ridges (Duane et al., 1972). As mentioned previously, the *upper ridge sand* unit volumetrically occupies the main portion of the active ridge. Its basal contact occurs along the flanks of the ridge so that the lower units are exposed below it.

The dynamics of ridge formation are still not fully understood. Huthnance's model (1982a, b), based on tidally formed ridges in the southern bight of the North Sea, is the best available. The North Sea ridges are very similar to the storm-built ridges of the New Jersey shelf, and their sediment dynamics are probably similar (J. Huthnance, pers. comm.). The Huthnance model is explained in Figure 12. Parker et al. (1982) offer a geologically oriented explanation of the Huthnance model and apply it to the storm-built sand ridges of the Argentina shelf.

The nearshore ridges on the Atlantic shelf originated on the shoreface in response to winter storm flows (Swift, 1976). These ridges are built with sand scoured from the shoreface as it undergoes erosional retreat in response to Holocene sea level rise. As the shoreface erodes out from under the ridges, they are lowered onto the shelf (Fig. 13).

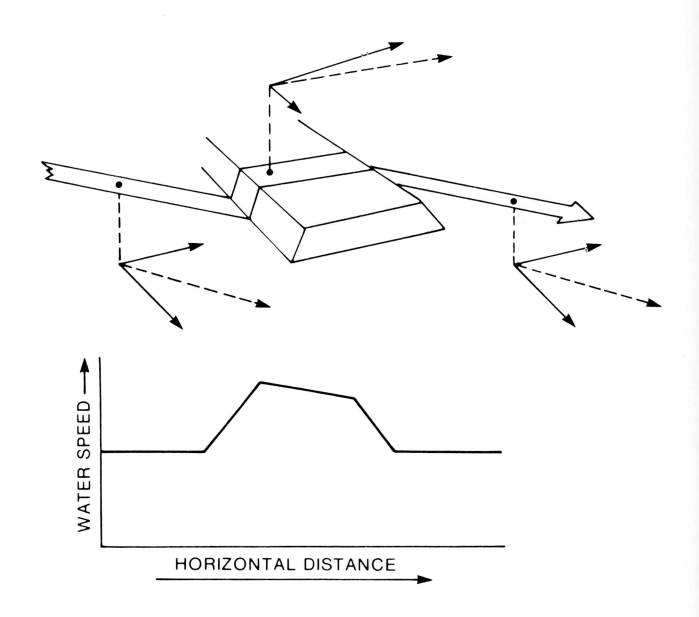

Figure 12. Huthnance model for the formation of a large-scale, flow-oblique sand ridge. As flow moves obliquely up the ridge flank, the cross-ridge component of velocity accelerates owing to decreasing cross-sectional area. Over the crest, the along-ridge component of velocity is reduced by frictional drag. The cross-ridge velocity component decreases to original values as flow passes over the down-current flank. Changed values for two flow components cause the current to veer as it passes over the ridge.

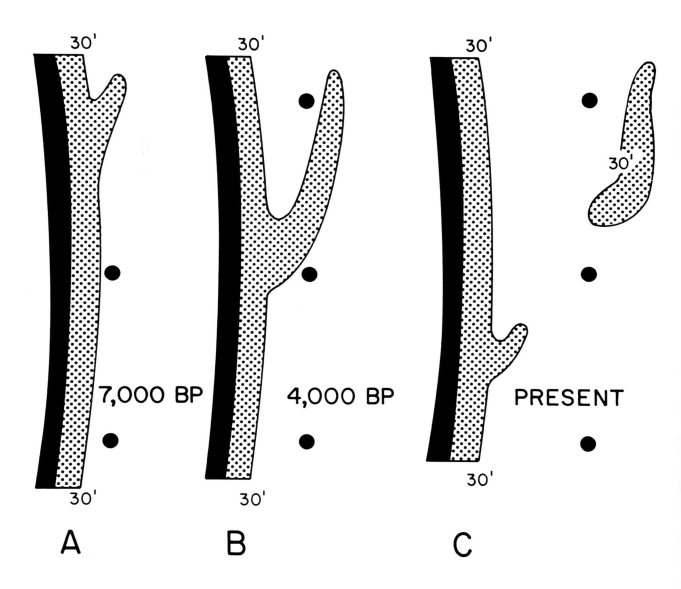

Figure 13. Schematic diagram illustrating the shoreface-detachment mode of nearshore sand ridge formation. Dots are fixed points with which to contrast respective positions of retreating shoreline and migrating ridges (from Swift, 1976).

Genesis of the Midshelf Ridge

Several hypotheses have been posited to explain the origin of the midshelf ridges on the Atlantic shelf. One describes the midshelf ridges as being cored by degraded barrier islands (Stubblefield et al., 1983; Stubblefield et al., 1984). A second model depicts the midshelf ridges as being formed on the leading edge of the shoal-retreat massif as it accumulated in shallow water and arrived at its midshelf position as the transgression continued (Swift et al., 1978; Swift et al., 1984).

Stubblefield et al. (1984) theorize that the midshelf ridges are "degraded" barriers that were submerged by rising sea level. This theory is based on the following characteristics of midshelf ridges: (1) they are oriented parallel to the present-day coastline; (2) they contain some fine- to medium-grained sands that appear to have been deposited in a relatively low energy setting and that are capped with 1 m of sand similar to nearshore ridge sands; and (3) they contain interlayered sands and muds. Figure 14 shows a schematic representation of the degraded barrier model by Stubblefield et al. (1984).

Swift et al. (1984) argue against a barrier overstep origin. They agree that the midshelf scarp represents a brief progradation of the shoreline during a general transgression and refer to this process as barrier "step-up," to be distinguished from true "overstep," where full barrier morphology is preserved. In barrier step-up, resumption of erosional shoreface retreat destroys barrier morphology; only a fossil lower shoreface is left as a scarp. Swift et al. (1984) conclude that the ridges landward of the scarp were formed by marine processes after the shoreline had passed.

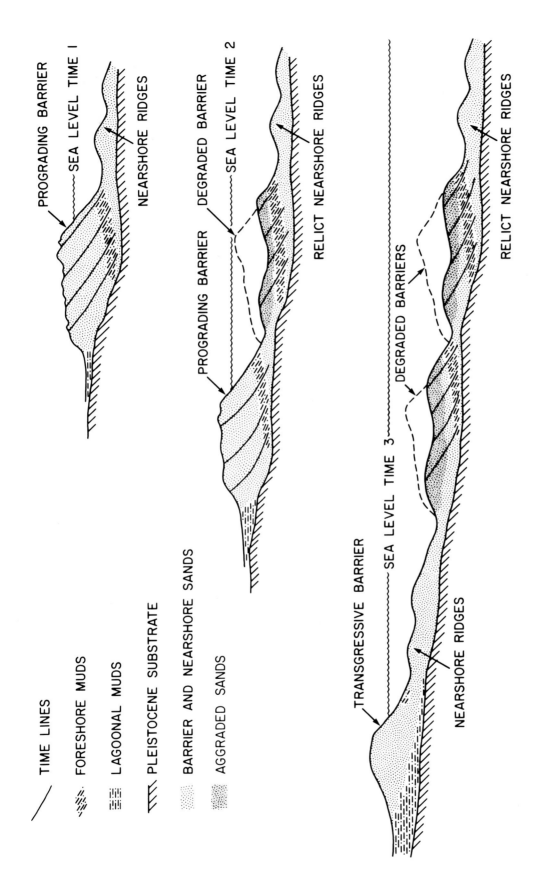

TIME LINES

FORESHORE MUDS

LAGOONAL MUDS

PLEISTOCENE SUBSTRATE

BARRIER AND NEARSHORE SANDS

AGGRADED SANDS

PROGRADING BARRIER

SEA LEVEL TIME 1

NEARSHORE RIDGES

DEGRADED BARRIER

SEA LEVEL TIME 2

RELICT NEARSHORE RIDGES

PROGRADING BARRIER

TRANSGRESSIVE BARRIER

SEA LEVEL TIME 3

DEGRADED BARRIERS

NEARSHORE RIDGES

DEGRADED BARRIERS

RELICT NEARSHORE RIDGES

Figure 14. Schematic diagram showing development of New Jersey midshelf ridges. Figure is not drawn to scale; barriers are somewhat enlarged in order to emphasize various features (from Stubblefield et al., 1984).

36

We (Rine, Tillman, and Swift) contend that the midshelf ridges actually formed in an offshore location under present shelf conditions (Fig. 15). Major evidence for this conclusion is that the *upper ridge sand* unit in the midshelf ridge appears to be one continuous unit having only gradational changes from top to bottom. There is a thin veneer (1-9 in.; 2.5-22 cm) of coarser-grained sand capping the ridge, but this coarse sand bed is also present on the nearshore ridge and can be attributed to normal fair-weather reworking (Stubblefield and Swift, 1970). If we acknowledge that the present top of the *upper ridge sand* unit is being deposited in a midshelf setting and that there are no significant breaks within the unit, then we must assume that uniform depositional conditions were present throughout deposition of the whole *upper ridge sand*. If this assumption is correct, then within some midshelf ridges as much as 18 ft (5.5 m) of sand has been deposited in a midshelf setting (Fig. 7, V-7A).

Sedimentologic differences between the *upper ridge sand* unit in the two ridges can be attributed to differences in processes and intensities of processes occurring at different water depths on the shelf. Lower average energy levels may have been typical of the midshelf, resulting in finer mean grain size, lower abundance of physical sedimentary structures, and thicker bedding within the midshelf ridge.

Conclusions

Core data from this study revealed that the following three main lithologic units are present in both the nearshore ridge and midshelf ridge: (1) the *"nonfossiliferous" mud and sand* unit deposited during the early to mid-Holocene or late Pleistocene; (2) the *shell-rich poorly sorted mud and sand* unit deposited as posttransgressive nearshore or

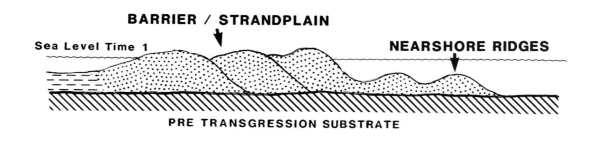

BARRIER / STRANDPLAIN

Sea Level Time 1

NEARSHORE RIDGES

PRE TRANSGRESSION SUBSTRATE

Sea Level Time 2

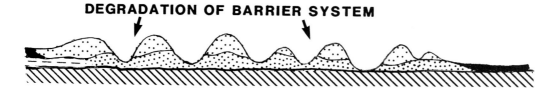

DEGRADATION OF BARRIER SYSTEM

Sea Level Time 3

SANDS REWORKED INTO MID-SHELF RIDGES

LEGEND

—— TIME LINES

‐‐‐ LAGOONAL MUDS

PRE TRANSGRESSION
SUBSTRATE

BARRIER AND NEARSHORE
SANDS

AGGRADED SHELF SANDS

SHELF SANDS AND MUDS

Figure 15. Schematic diagram of a model for posttransgressive development of midshelf ridges with major modification occurring in the present-day midshelf setting. Figure is not drawn to scale.

shelf sediments; and (3) the *upper ridge sand* unit, which is the actively aggrading ridge-forming unit within both ridges.

Core data from this study supports a nearshore, shoreface-attached origin during late Holocene time for the nearshore ridge at location 1A. Core data from this study indicate that the midshelf ridges were all or partly formed at the present stand of sea level.

Acknowledgments

The authors thank the companies belonging to the Atlantic Shelf Coring Project consortium for releasing the data presented in this paper. Companies that have participated in the consortium are Atlantic Richfield Oil and Gas (ARCO), Amoco Production Company, Getty Oil Company, SOHIO Oil and Gas Company, Cities Service Oil and Gas Corporation, Mobil Oil and Gas Company, and Chevron Oil and Gas Company. The study was a cooperative project, undertaken by the consortium and by the Atlantic Oceanographic and Meteorological Laboratories, National Oceanic and Atmospheric Administration.

Rod Tillman and Bill Stubblefield acted as co-chief scientists on the 1979 Atlantic Shelf Project cruise, which was carried out on the *R.V. Eastward*. Jim Rine supervised the vibracore operations; he was assisted by Chuck Dill and Norman Haskell. Additional scientific crew members included Chuck Siemers, Randi Martinsen, Ted Beaumont, and Dick Scott. Fred Mason and Chris Berlin were on-board technicians.

The crew of the *R.V. Eastward* did a good job of supporting the scientific crew in spite of interruption by tropical storm David. The vibracores were processed at the Cities Service Research facility in Tulsa, Oklahoma, under supervision of Jim Rine and Rod Tillman; they were assisted by Fred Mason, Jim Pol, Pierre Kinga, and Chris Berlin. Typing and drafting support was provided by Cities Service.

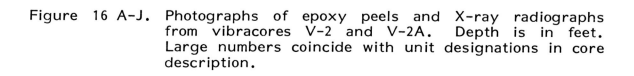

Figure 16 A-J. Photographs of epoxy peels and X-ray radiographs from vibracores V-2 and V-2A. Depth is in feet. Large numbers coincide with unit designations in core description.

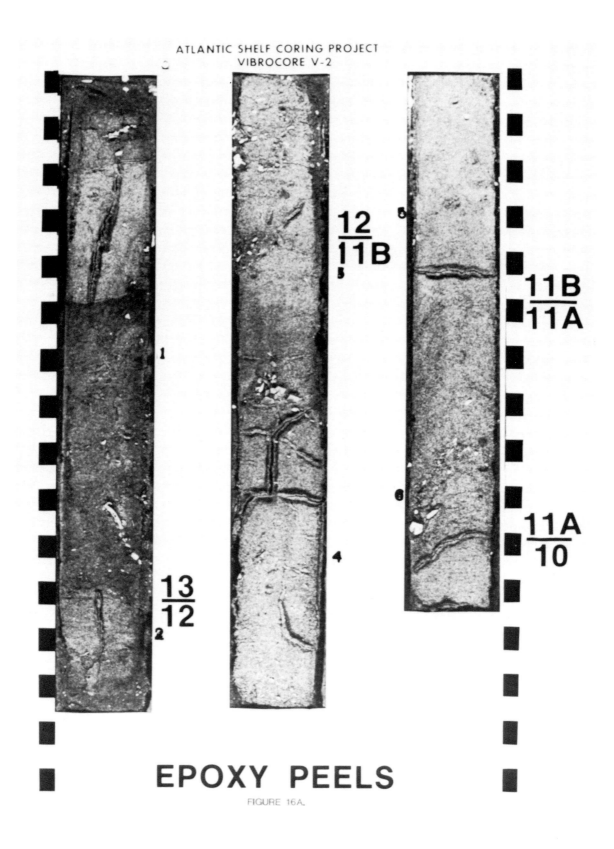

EPOXY PEELS

FIGURE 16A.

Figure 16 A-J. Photographs of epoxy peels and X-ray radiographs from vibracores V-2 and V-2A. Depth is in feet. Large numbers coincide with unit designations in core description.

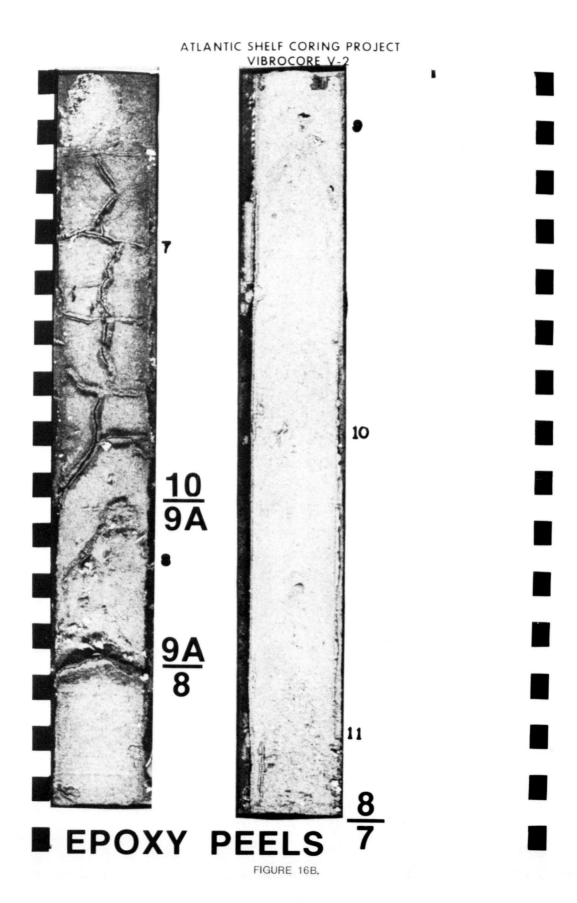

EPOXY PEELS

FIGURE 16B.

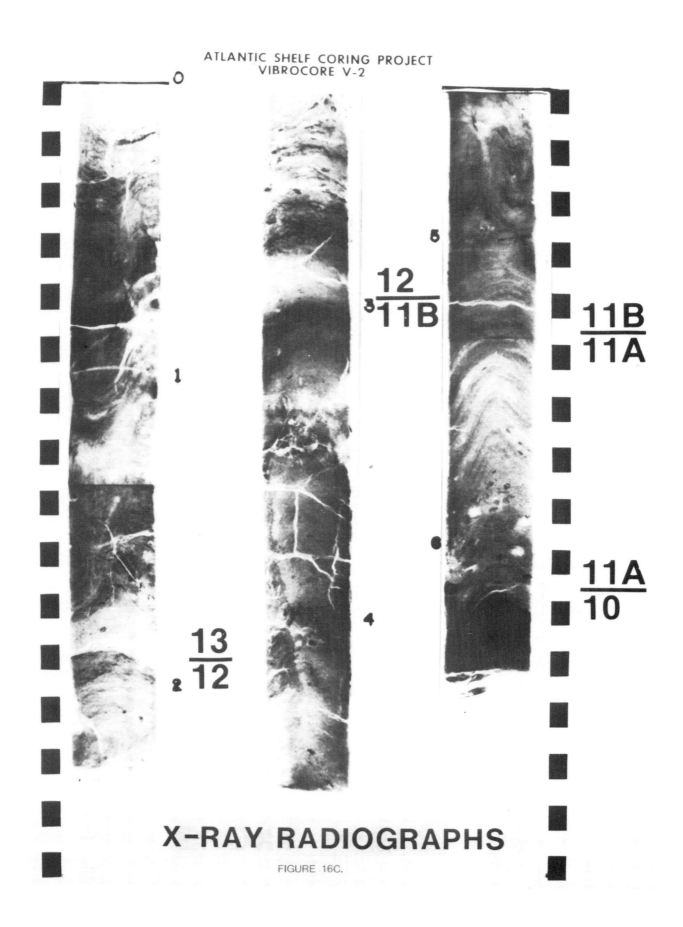

ATLANTIC SHELF CORING PROJECT
VIBROCORE V-2

X-RAY RADIOGRAPHS

FIGURE 16C.

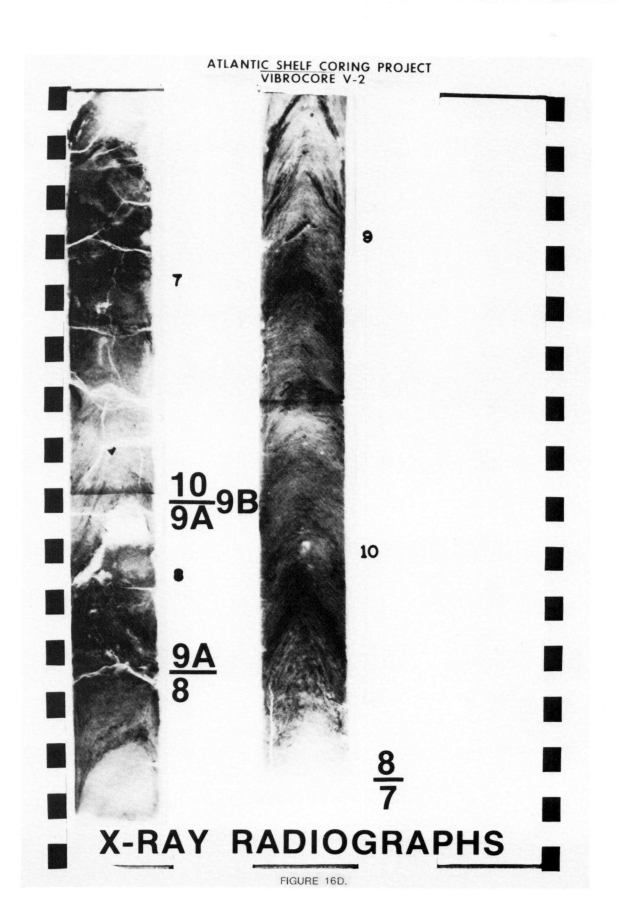

X-RAY RADIOGRAPHS

FIGURE 16D.

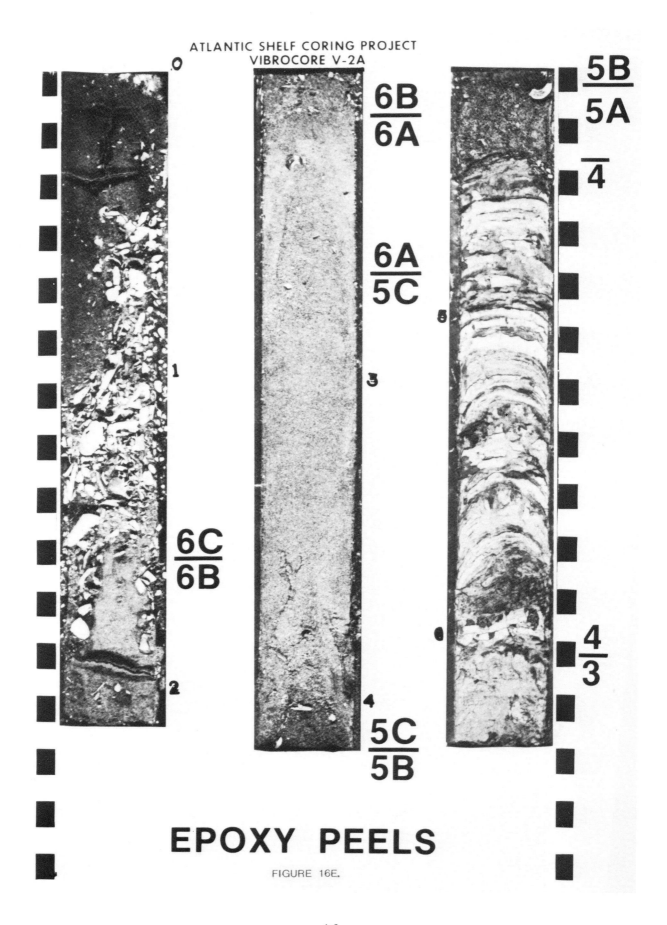

ATLANTIC SHELF CORING PROJECT
VIBROCORE V-2A

6B/6A

6A/5C

6C/6B

5C/5B

5B/5A

4

4/3

EPOXY PEELS

FIGURE 16E.

46

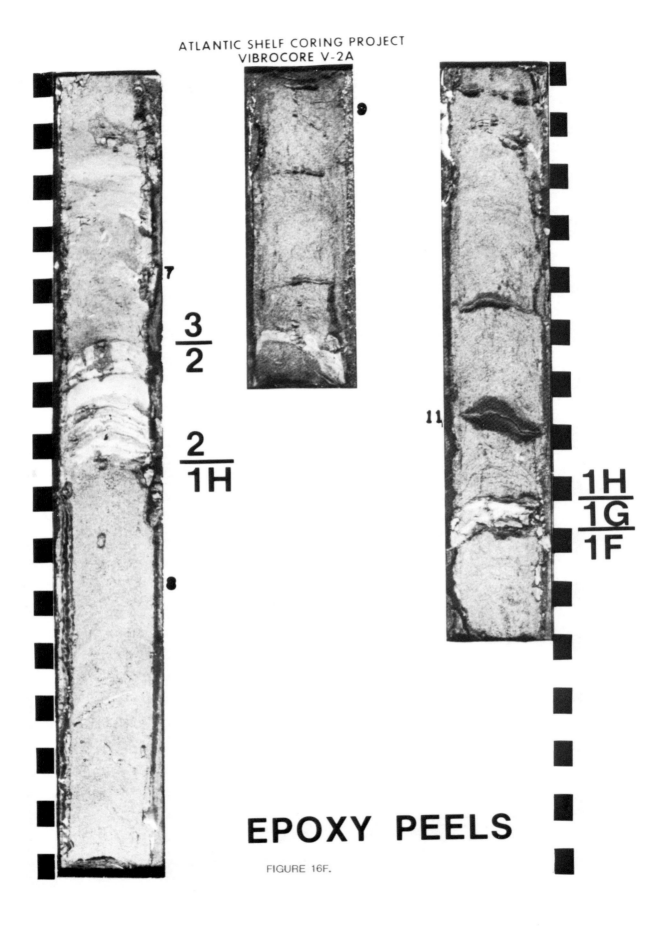

ATLANTIC SHELF CORING PROJECT
VIBROCORE V-2A

$\dfrac{3}{2}$

$\dfrac{2}{1H}$

$\dfrac{1H}{1G}$
$\overline{}$
$1F$

EPOXY PEELS

FIGURE 16F.

12 1F
1E
1D

1D
1C

13

1C
1B
1A

EPOXY PEELS

FIGURE 16G.

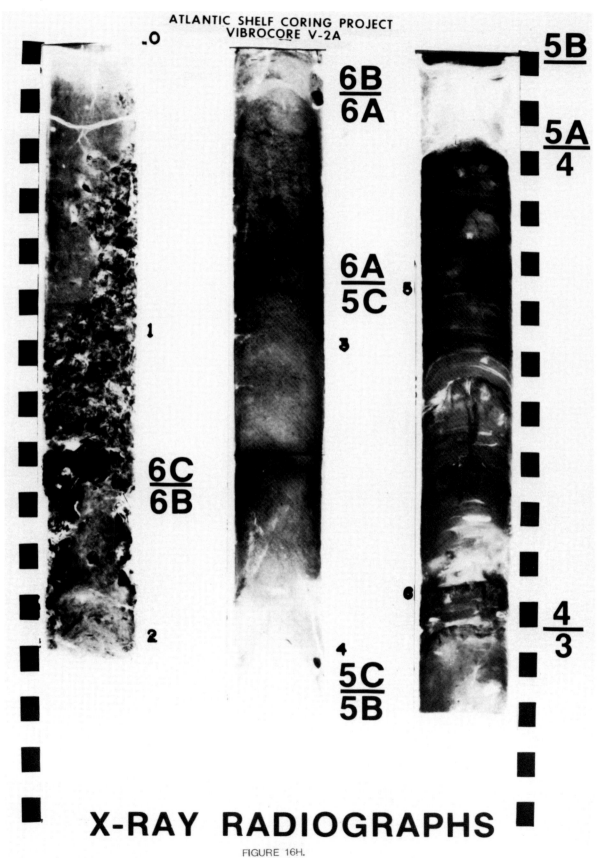

ATLANTIC SHELF CORING PROJECT
VIBROCORE V-2A

X-RAY RADIOGRAPHS

FIGURE 16H.

7

$$\frac{3}{2}$$

$$\frac{2}{1H}$$

9

10

11

$$\frac{1H}{1G}$$
$$\frac{1G}{1F}$$

8

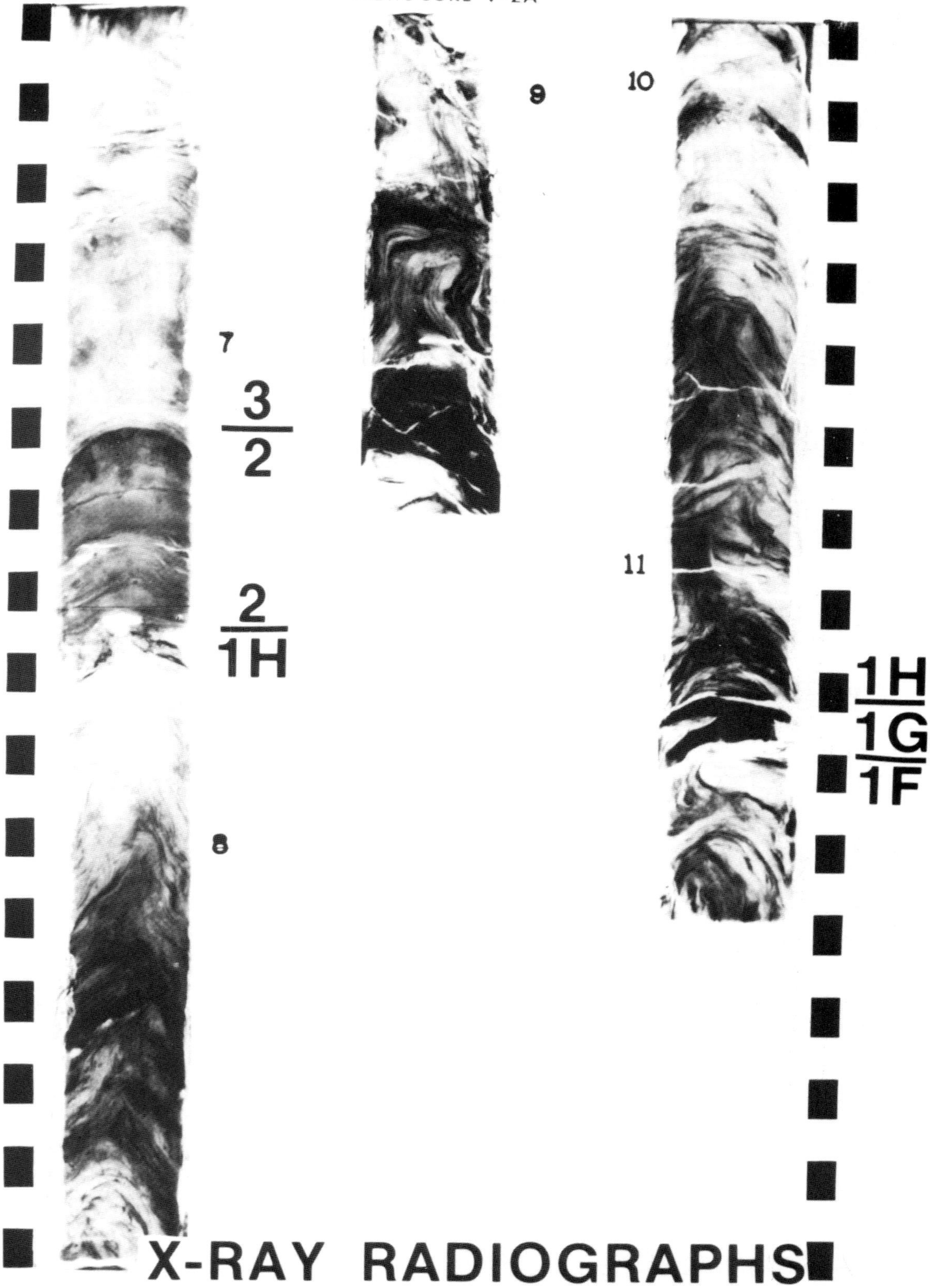

X-RAY RADIOGRAPHS

FIGURE 16L

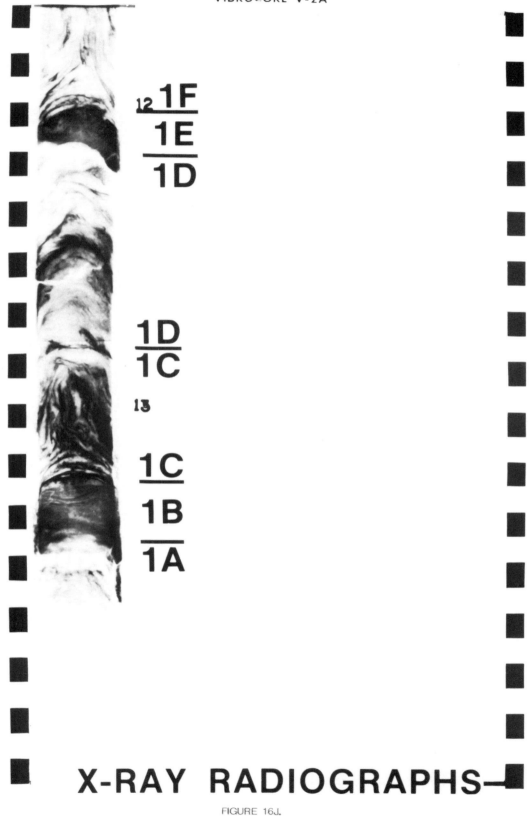

X-RAY RADIOGRAPHS

FIGURE 16J.

Figure 17 A-F. Photographs of epoxy peels and X-ray radiographs from vibracore V-10. Depth is in feet. Large numbers coincide with unit designations in core description.

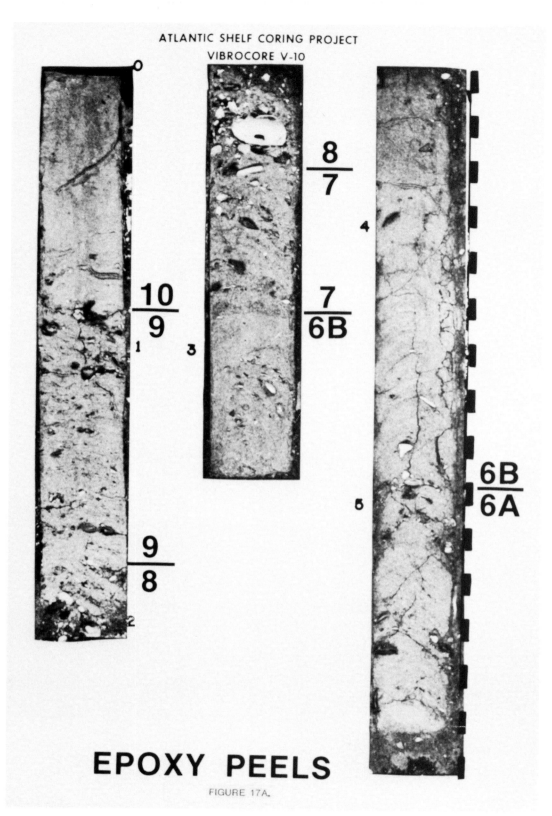

ATLANTIC SHELF CORING PROJECT

VIBROCORE V-10

EPOXY PEELS

FIGURE 17A.

Figure 17 A-F. Photographs of epoxy peels and X-ray radiographs from vibracore V-10. Depth is in feet. Large numbers coincide with unit designations in core description.

ATLANTIC SHELF CORING PROJECT
VIBROCORE V-10

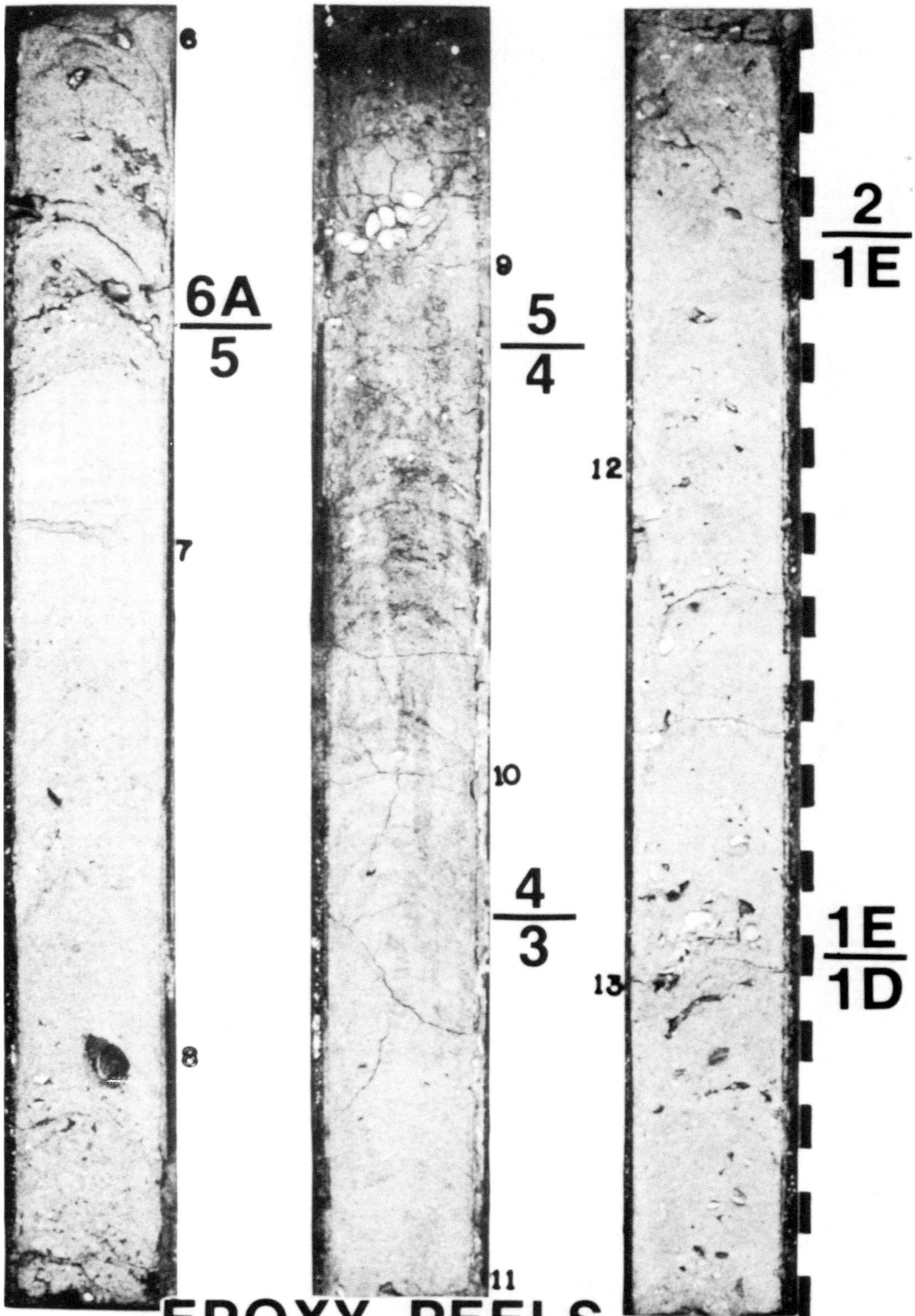

EPOXY PEELS

FIGURE 17B.

55

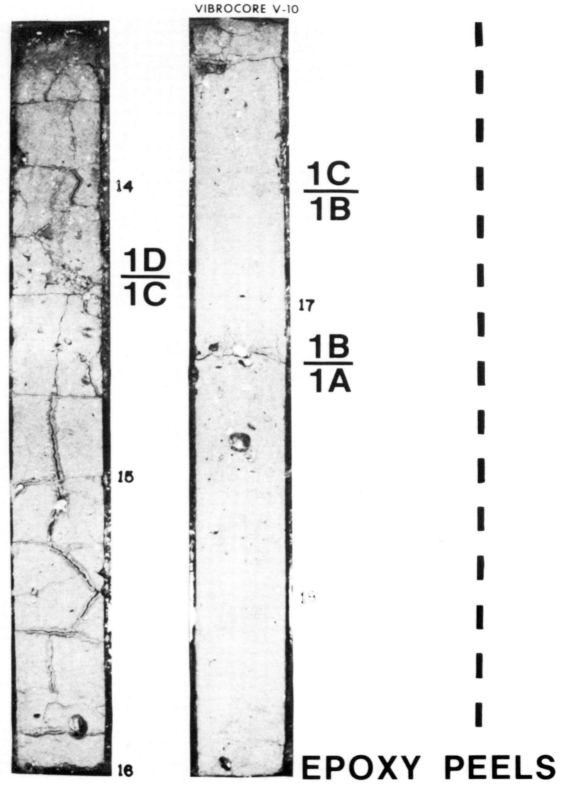

EPOXY PEELS

FIGURE 17C.

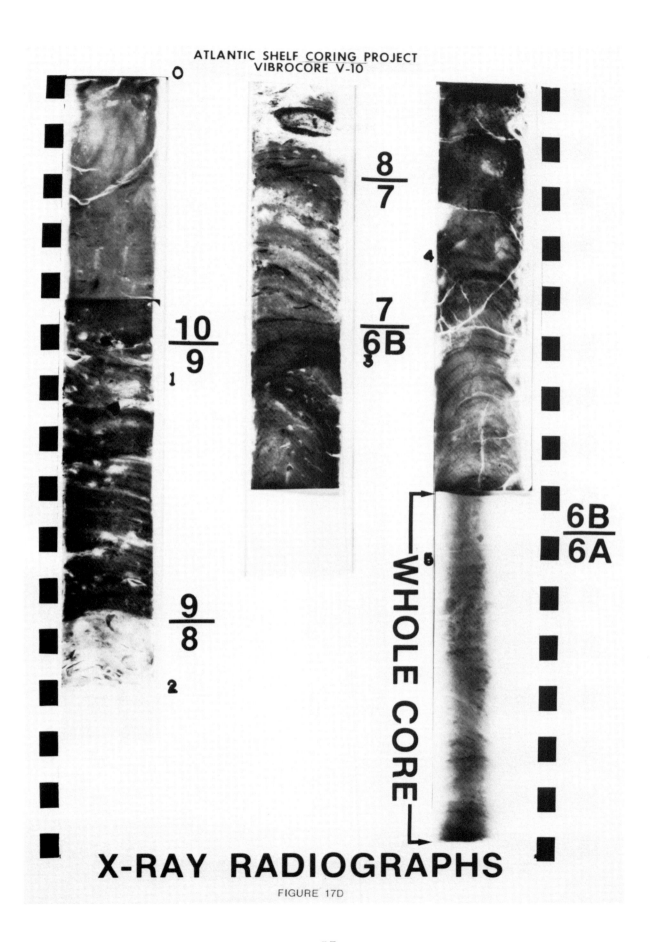

X-RAY RADIOGRAPHS

FIGURE 17D

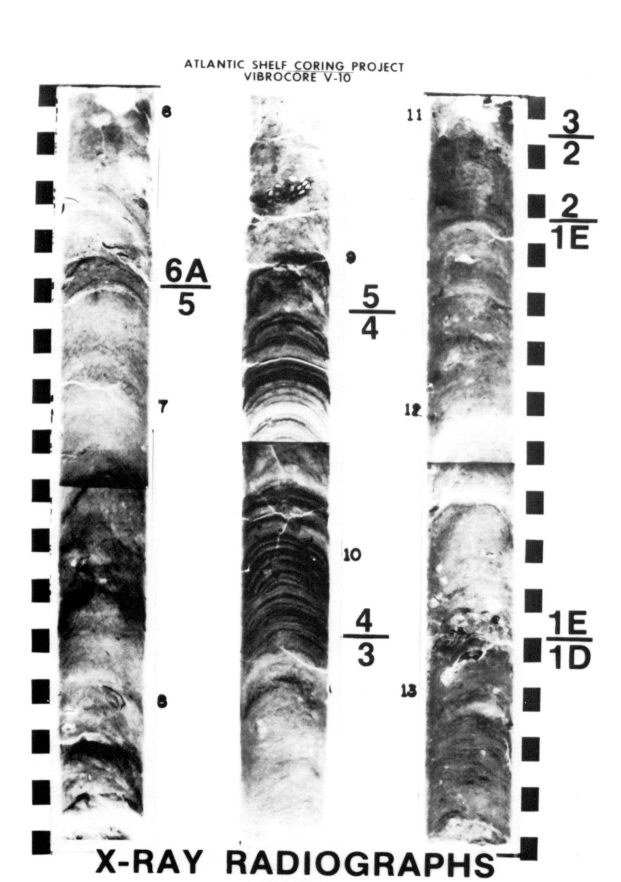

X-RAY RADIOGRAPHS

FIGURE 17E

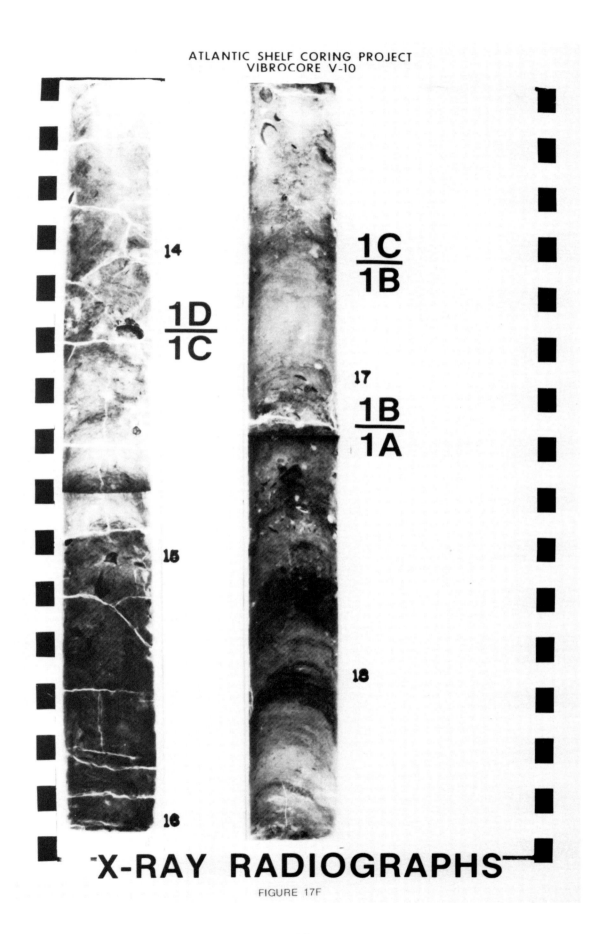

X-RAY RADIOGRAPHS

FIGURE 17F

Figure 18 A-F. Photographs of epoxy peels and X-ray radiographs from vibracore V-13. Depth is in feet. Large numbers coincide with unit designations in core description.

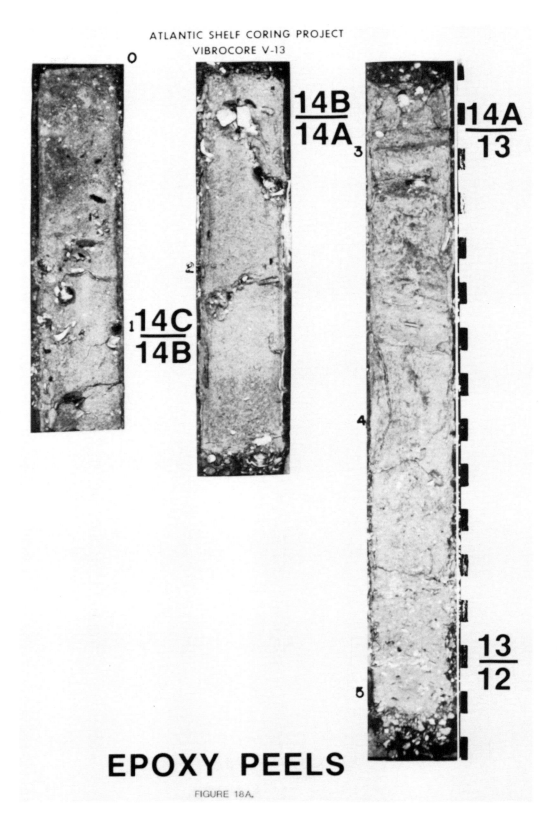

ATLANTIC SHELF CORING PROJECT
VIBROCORE V-13

EPOXY PEELS

FIGURE 18A.

Figure 18 A-F. Photographs of epoxy peels and X-ray radiographs from vibracore V-13. Depth is in feet. Large numbers coincide with unit designations in core description.

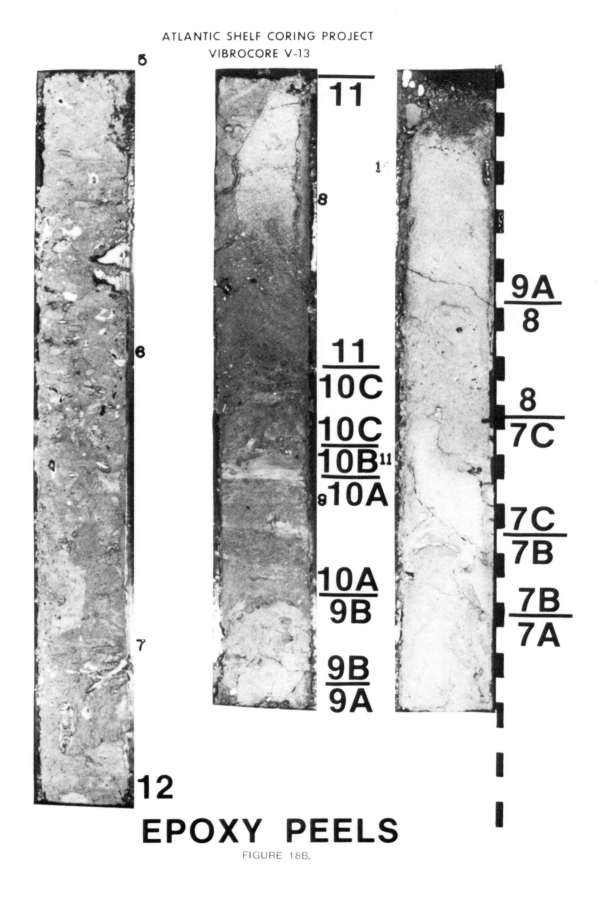

ATLANTIC SHELF CORING PROJECT
VIBROCORE V-13

EPOXY PEELS

FIGURE 18B.

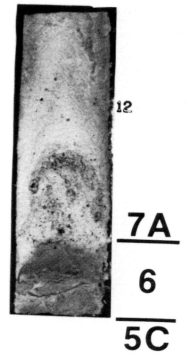

12

7A

6

5C

EPOXY PEELS

FIGURE 18C.

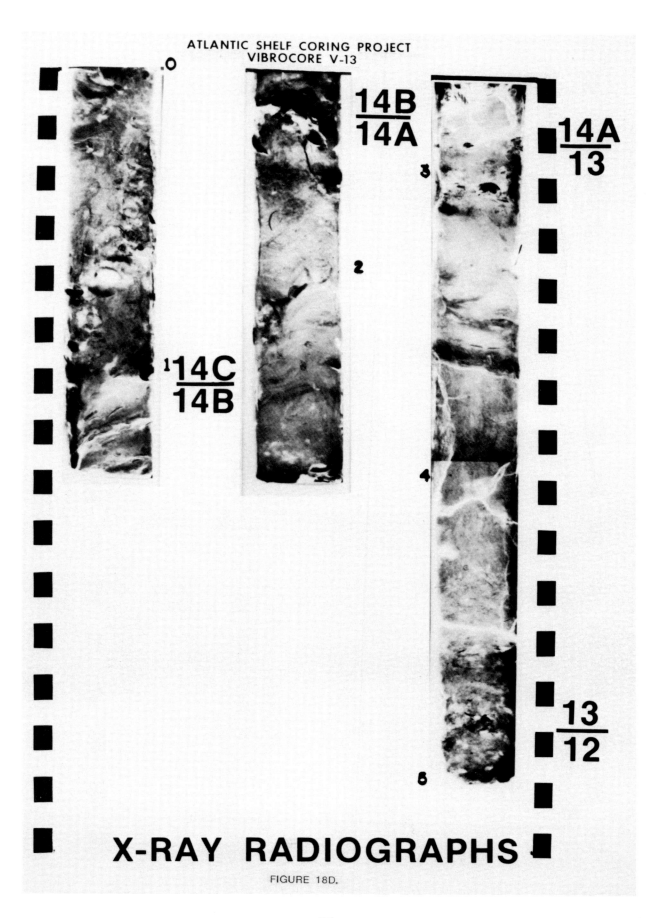

14B
14A

14A
13

¹14C
14B

13
12

X-RAY RADIOGRAPHS

FIGURE 18D.

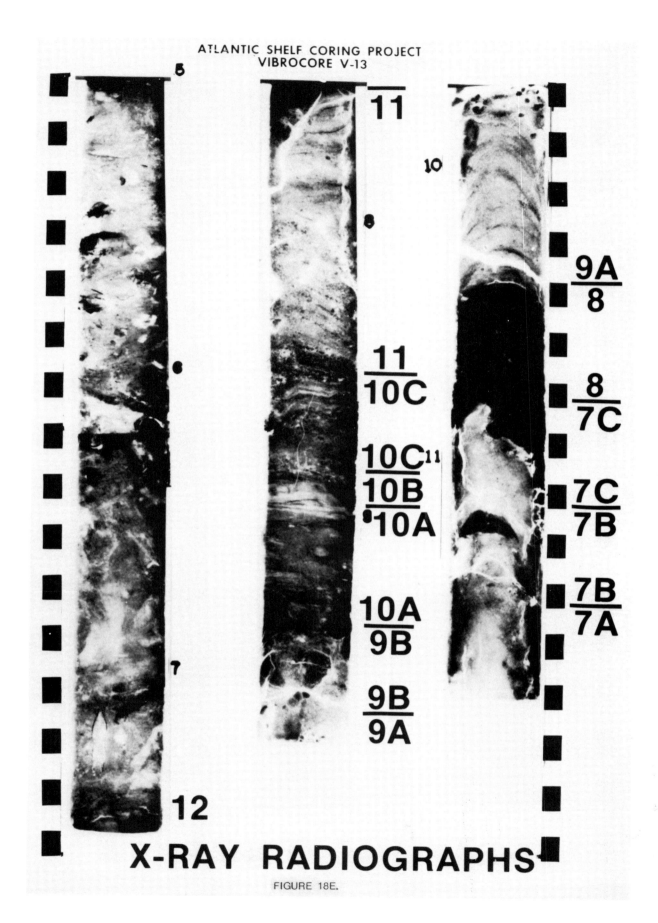

X-RAY RADIOGRAPHS

FIGURE 18E.

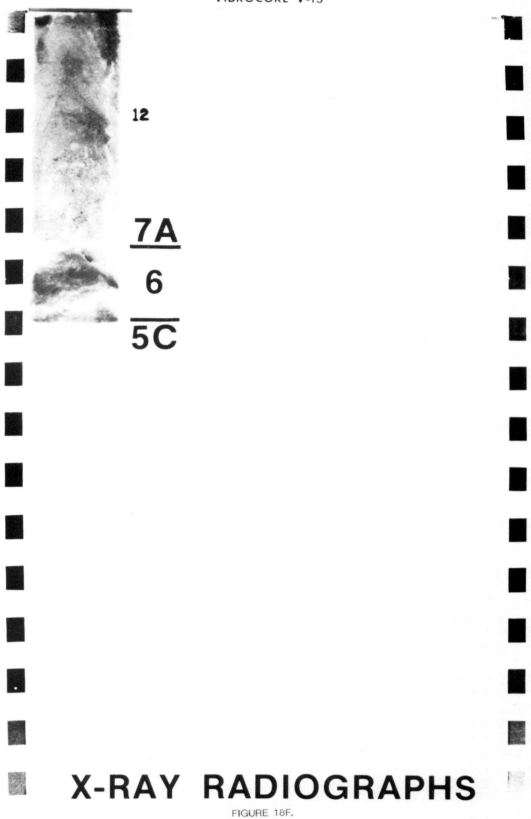

12

7A

6

5C

X-RAY RADIOGRAPHS

FIGURE 18F.

References

Butman, B., Noble, M., and Folger, D. W., 1979, Long-term observations of bottom current and bottom sediment movement on the mid-Atlantic continental shelf: Jour. Geophys. Res., v. 84, no. C3, p. 1187-205.

Duane, D. B., Field, M. E., Miesberger, E. P., Swift, D. J. P., and Williams, S. J., 1972, Linear shoals on the Atlantic continental shelf, Florida to Long Island, in Swift, D. J. P., Duane, D. B., and Pilkey, D. H., eds., Shelf sediment transport, process and pattern: Stroudsburg, Pa., Dowden and Ross, p. 447-98.

Emery, K. O., 1952, Continental shelf sediments of southern California: Bull. Geol. Soc. Amer., v. 63, p. 1105-8.

Emery, K. O., 1965, Geology of the continental margin off eastern United States, in Whittard, W. F., and Bradshaw, R., eds., Submarine geology and geophysics: London, Bulter Worths, p. 1-20.

Harrison, W., Norcross, J. J., Pore, N. A., and Stanley, E. M., 1967, Shelf waters off the Chesapeake Bight: Environmental Sciences Services Administration, Prof. Paper 3, p. 1-82.

Howard, J. D., 1972, Trace fossils as criteria for recognizing shorelines in the stratigraphic record, in Rigby, J. D., and Hamblin, W. K., eds., Recognition of ancient sedimentary environments: Soc. Econ. Pal. Min. Spec. Publ. No. 16, p. 215-55.

Howard, J. D., and Reineck, H. E., 1981, Depositional facies of high energy beach-to-offshore sequence--comparison with low-energy sequence: Amer. Assoc. Petrol. Geol. Bull., v. 65, p. 807-30.

Huthnance, J. M., 1982a, On one mechanism forming linear sand banks: Estuarine Coastal Shelf Sci., v. 14, p. 79-99.

Huthnance, J. M., 1982b, On formation of sand banks of definite extent: Estuarine Coastal Shelf Sci., v. 15, p. 277-99.

Knebel, H. J., and Spiker, E., 1977, Thickness and age of surficial sand sheet, Baltimore Canyon Trough area: Amer. Assoc. Petrol. Geol. Bull., v. 61, p. 861-71.

Moody, D. W., 1964, Coastal morphology and processes in relation to the development of submarine sand ridges off Bethany Beach, Delaware: Ph.D. dissertation, Johns Hopkins Univ., Baltimore, Maryland, 167 p.

Moore, D. G., and Curray, J. R., 1974, Midplate continental margin geosynclinal growth processes and Quaternary modifications: Soc. Econ. Pal. Min. Spec. Publ. No. 19, p. 26-35.

Parker, G., Landfredi, N. W., and Swift, D. J. P., 1982, Substrate response flow in a southern hemisphere ridge field, Argentina Inner Shelf: Sediment. Geol., v. 33, p. 195-216.

Shepard, F. P., and Cohee, G. V., 1936, Continental shelf sediments off the mid-Atlantic states: Bull. Geol. Soc. Amer., v. 47, p. 441-58.

Stetson, H. C., 1938, The sediments of the continental shelf off the eastern Coast of the United States: MIT and WHOI Papers in Phys. Oceanogr. and Meteorol., v. 5, no. 4, 48 p.

Stubblefield, W. L., Kersey, D. G., and McGrail, D. W., 1983, Development of middle continental shelf sand ridges, New Jersey: Amer. Assoc. Petrol. Geol. Bull., v. 67, p. 817-30.

Stubblefield, W. L., McGrail, D. W., and Kersey, D. G., 1984, Recognition of transgressive and post-transgressive sand ridges on the New Jersey continental shelf, in Tillman, D. W., and Siemers,

C. T., eds., Siliciclastic shelf sediments: Soc. Econ. Pal. Min. Spec. Publ. No. 34, p. 1-23.

Stubblefield, W. L., and Swift, D. J. P., 1976, Ridge development as revealed by subbottom profiles on the central New Jersey shelf: Marine Geol., v. 20, p. 315-34.

Swift, D. J. P., 1973, Delaware Shelf Valley--estuary retreat path, not drowned river valley: Geol. Soc. Amer. Bull., v. 84, p. 2743-248.

Swift, D. J. P., 1976, Continental shelf sedimentation, in Stanley, D. J., and Swift, D. J. P., eds., Marine sediment transport and environmental management: John Wiley and Sons, Inc., New York, p. 311-50.

Swift, D. J. P., Duane, D. B., and McKinney, T. F., 1973, Ridge and swale topography of the Middle Atlantic Bight, North America-- secular response to the Holocene hydraulic regime: Marine Geol., v. 15, p. 227-47.

Swift, D. J. P., and Field, M., 1981, Evolution of a classic sand ridge field, Maryland sector, North American inner shelf: Sedimentology, v. 28, p. 461-82.

Swift, D. J. P., Holliday, B., Avignone, N., and Shideler, G., 1972, Anatomy of a shoreface ridge system, False Cape, Virginia: Marine Geol., v. 12, p. 59-84.

Swift, D. J. P., Kofoed, J. W., Saulsbury, F. P., and Sears, P., 1972, Holocene evolution of the shelf surface, south and central Atlantic shelf of North America, in Swift, D. J. P., Duane, D. B., and Pilkey, O. H., eds., Shelf sediment transport, process and pattern: Stroudsberg, Pa., Dowden, Hutchinson, and Ross, p. 499-574.

Swift, D. J. P., McKinney, T. F., and Stahl, L., 1984, Recognition of transgressive and post-transgressive sand ridges on the New Jersey continental shelf, discussion, in Tillman, R. W., and Siemers, C. T., Siliclastic shelf sediments: Soc. Econ. Pal. Min. Spec. Publ. No. 34, p. 25-36.

Swift, D. J. P., and Sears, P., 1974, Estuarine and littoral deposition patterns in the surficial sand sheet, central and southern Atlantic shelf of North America: Mem. 7, Inst. Geol. Bassin Aquitaine, 1974, p. 171-89.

Swift, D. J. P., Sears, P. C., Bohlke, B., and Hunt, R., 1978, Evolution of a shoal retreat massif, North Carolina shelf--inferences from areal geology: Marine Geol., v. 27, p. 19-47.

INNER-SHELF SHOAL SEDIMENTARY FACIES AND SEQUENCES:

SHIP SHOAL, NORTHERN GULF OF MEXICO

Shea Penland, John R. Suter, and Thomas F. Moslow

Louisiana Geological Survey, Coastal Geology Program, University Station, Box G, Baton Rouge, Louisiana 70893

Abstract

Shore-parallel sand shoals are common depositional features on the continental shelves of North America and are a significant component of the stratigraphic record generated by the Holocene transgression. *Erosional shoreface retreat* and *in-place drowning* of barrier shorelines have been incorporated in models to explain the deposition and evolution of these features. Sand shoals within the retreat path of the Holocene Mississippi River deltas were investigated using high-resolution seismic profiles and vibracores to determine their stratigraphic development and to test the models proposed for inner-shelf shoal formation.

Ship Shoal is a shore-parallel sand body 50 km long having a relief of 3-6 m. Historic records suggest that it is migrating landward 10-15 m/yr and is presently located 25 km from the shoreline in water depths of 7-15 m. The Ship Shoal sedimentary sequence is 4-6 m thick and coarsens upward through the shoal-base, shoal-front, and shoal-crest facies. The shoal lies disconformably over deposits of the Maringouin delta, which was transgressed approximately 4000 years BP. A laterally continuous unit of lagoonal muds at the base of the shoal sequence attests to the former existence of barrier shoreline environments. No in-situ barrier shoreline deposits were found within the shoal sand body.

Data indicate that Ship Shoal is a marine sand body sourced from the erosion of a former barrier shoreline through a process termed *transgressive submergence*, which integrates the mechanisms of shoreface erosion and relative sea level rise within the process of coastal submergence. The models of *erosional shoreface retreat* and *in-place drowning* do not adequately explain the morphology or stratigraphy of Mississippi Delta inner-shelf shoals. Ship Shoal can be used as a model to explain the occurrence of shelf sandstones observed tens of kilometers from the penecontemporaneous shoreline in interior seaways and foreland basins.

Introduction

Ship Shoal is the easternmost member of a group of Holocene inner-shelf shoals located in Louisiana offshore of the southwestern Mississippi River delta plain (Fig. 1). Ship Shoal is a shore-parallel

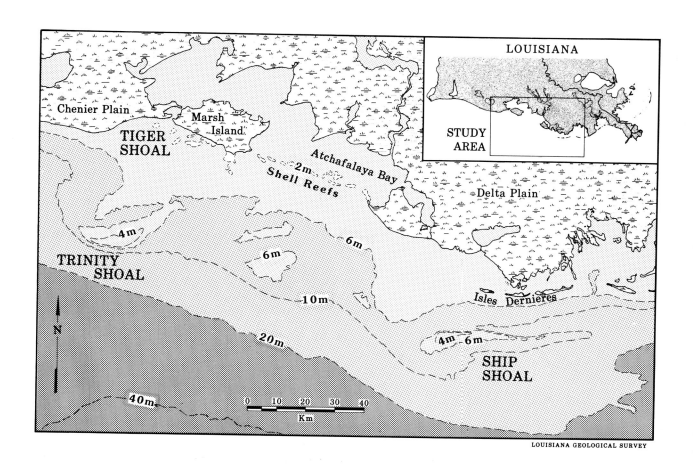

Figure 1. Ship Shoal and Trinity Shoal lie offshore of the southwestern Mississippi River delta plain in Louisiana.

sand body 50 km long and 8-12 km wide lying in 10 m of water; it has a relief above the surrounding shelf of 3-7 m east to west along its crest axis. A comparison of hydrographic surveys indicates that the shoal has migrated 1.5 km landward since 1850. On the basis of shelf morphology, lithology, bathymetric changes, and surface texture, our investigations document that Ship Shoal is a marine sand body derived from the erosion of a submerged barrier shoreline (Fig. 2). The transgressive stratigraphic signature of the Ship Shoal sequence illustrates a process termed *transgressive submergence* (Penland et al., in press) which integrates the mechanisms of shoreface erosion and relative sea level rise during coastal submergence. Our objective in this core workshop manuscript is to describe the sedimentary facies and development of Ship Shoal. We will compare and contrast Ship Shoal with other well-documented continental shelf sand bodies, such as the U.S. Atlantic shelf sand ridges, and examine the stratigraphic record generated during relative sea level rise.

The data base used in this analysis consists of more than 1000 km of single-channel, high-resolution seismic profiles (Datasonics 3.5 kHz subbottom profiler and ORE Geopulse systems), 40 offshore vibracores (10-14 m x 10 cm), more than 50 borings, and more than 100 vibracores (5-9 m x 7 cm) on the immediate mainland. Figure 2 depicts the data base in the immediate Ship Shoal region, as well as locations of geologic cross sections.

Regional Geology

The Mississippi River has built its delta plain through a sequential process of delta-building episodes followed by abandonment and transgressive barrier shoreline generation. The evolutionary process is

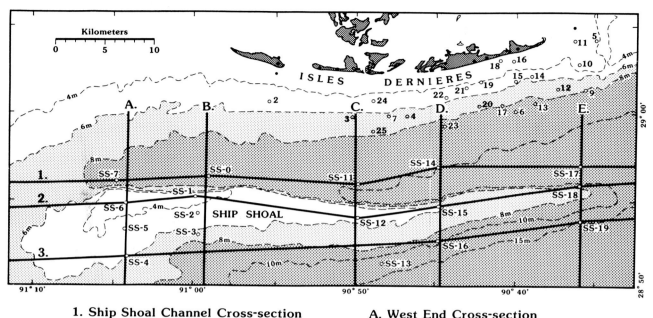

1. Ship Shoal Channel Cross-section
2. Ship Shoal Crest Cross-section
3. Ship Shoal Shelf Cross-section

A. West End Cross-section
B. Western Region Cross-section
C. Central Cross-section
D. Eastern Region Cross-section
E. East End Cross-section

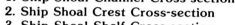

TEXTURE

:::: Medium Sand   Very Fine Sand

   Fine Sand   Silty Clay

SEDIMENTARY STRUCTURES

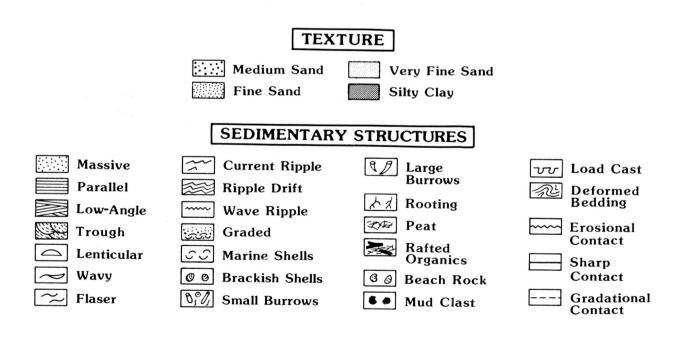

Massive	Current Ripple	Large Burrows	Load Cast
Parallel	Ripple Drift	Rooting	Deformed Bedding
Low-Angle	Wave Ripple	Peat	Erosional Contact
Trough	Graded	Rafted Organics	
Lenticular	Marine Shells	Beach Rock	Sharp Contact
Wavy	Brackish Shells	Mud Clast	Gradational Contact
Flaser	Small Burrows		

Figure 2. Location diagram of vibracores and cross sections in the Ship Shoal region. The texture and sedimentary structure legends for all cored sequences are listed below.

collectively known as the delta cycle (Fisk, 1944; Kolb and Van Lopik, 1958; Scruton, 1960; Coleman and Gagliano, 1964; Frazier, 1967). The morphology of the Mississippi River delta plain and inner continental shelf reflects the combined effects of the regressive and transgressive phases of the delta cycle operating under the conditions of sea level rise. The Holocene delta plain consists of an extensive network of smaller deltas radiating seaward from the Mississippi alluvial valley into the Gulf of Mexico between Baton Rouge and Lafayette (Fig. 3). These deltas are separated by a series of connecting interdistributary lakes and ponds that increase in size and coalesce toward the coast, forming large bays open to the Gulf (Fisk, 1944; Gould, 1970). The Mississippi River has built six major delta complexes comprising more than 18 smaller deltas over the last 7000 years (Frazier, 1967).

Between 6000-7000 years BP and the present, delta building began in the area of the Isles Dernieres (Fig. 3). Ship Shoal is associated with the Maringouin delta complex, which was deposited when mean sea level was 5-8 m below the present level. Deltaic progradation began as early as 7900 years BP and continued until 6200 years BP (Frazier, 1967). The locus of Maringouin delta-complex sedimentation was concentrated immediately south of the Wisconsinan Mississippi alluvial valley on what is now the inner continental shelf. The rate of relative sea level rise increased after 4000 years BP, submerging the Maringouin delta complex. Ship Shoal is thought to have originated from the transgression of the Maringouin delta complex and to represent an ancient barrier shoreline (Krawiec, 1966; Frazier, 1967; Penland et al., 1981; Cuomo, 1984; Nummedal et al., 1984).

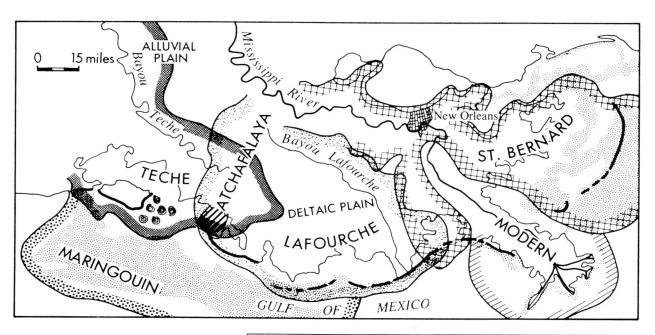

DELTA COMPLEX	AGE (YEARS BP)
Maringouin	7250-6200
Teche	5700-3900
St. Bernard	4600-1800
Lafourche	3500-400
Modern	Active
Atchafalaya	Active

⌒ Barrier Shoreline

Figure 3. The Holocene Mississippi River delta plain comprises six major delta complexes; two are currently active, the Atchafalaya and Modern, and four are abandoned, the Maringouin, Teche, St. Bernard, and Lafourche (from Frazier, 1967).

Inner-Shelf Morphology

The landward-oriented asymmetry of Ship Shoal suggests that it is migrating onshore in a north-northwest direction toward a bathymetric protuberance defined by the 6-m isobath that extends southeast out of the Caillou Bay and Point Au Fer area (Fig. 4A). Over the eastern end of the shoal, landward slopes of 1:750 and seaward slopes of 1:900 occur (Fig. 4A, B). Westward the landward slopes increase to 1:90, and seaward slopes decrease to 1:2100. The change in shoal-crest asymmetry and orientation westward is concurrent with a decrease in water depth over the crest and an increase in shoal relief.

A broad platform approximately 15-20 km wide lies seaward of the western half of Ship Shoal between the 18- and 20-m isobaths (Fig. 5). Average slope across this platform southward from Ship Shoal varies between 1:2000 and 1:3000. Approximately 25 km seaward of Ship Shoal on the seaward edge of this platform lies another shore-parallel shoal that has an inner-shelf relief of 1-2 m and water depths of 12-15 m above its crest. This feature is 35 km long and 5-7 km wide. The platform and outer shoal are best defined on the inner shelf seaward of Ship Shoal's western end. Westward, the slopes of the platform and outer shoal increase seaward, and the seaward slope of Ship Shoal decreases seaward; they merge to form a common slope of approximately 1:2200.

Shelf Surface Sediments

Krawiec (1966) examined the textural character and mineralogy of Ship Shoal and the adjacent shelf using grab samples from eastern and western shoal transects and concluded, on the basis of grain size, sorting, bulk composition, feldspar abundance, and heavy mineral

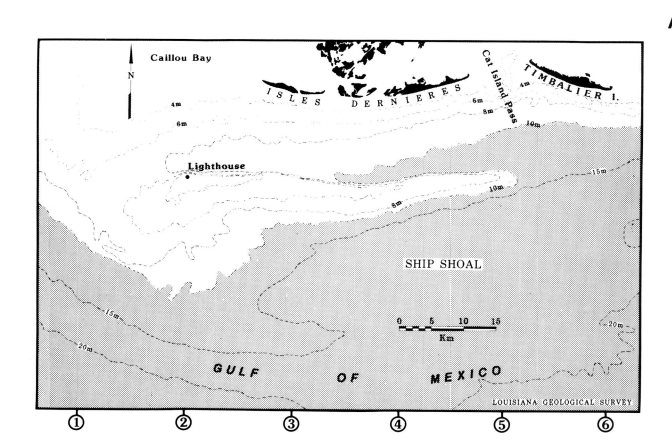

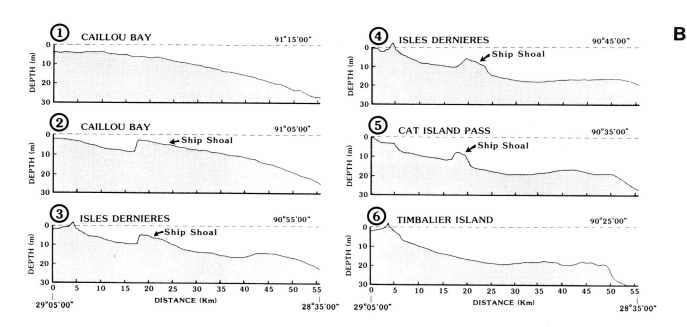

Figure 4. [A] Bathymetric diagram of the Ship Shoal and Isles Dernieres region. [B] Diagram of a series of evenly spaced bathymetric profiles extending offshore from the Lafourche delta complex shoreline and seaward across Ship Shoal. Profile locations are shown in Figure 4A.

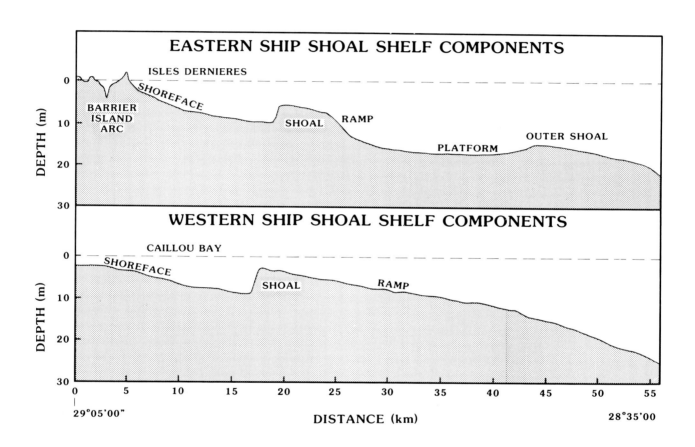

Figure 5. Two bathymetric profiles contrasting the major geomorphic components of the east and west Ship Shoal shelf region.

composition, that Ship Shoal was formed by processes that destroyed a barrier shoreline. Frazier (1974) mapped the subaqueous lithofacies in the Ship Shoal region on the basis of percentage of sand (Fig. 6); Ship Shoal is mapped as 75%-100% sand within the 7-8 m isobath. Where the isobaths protrude seaward in the western shoal region, inner-shelf sediments are 75%-100% sand; seaward, sediments become 50%-75% sand. The base of the Isles Dernieres shoreface surface is silty clay. The inner-shelf platform is also silty clay, and the crest of the outer shoal is 75%-100% sand. A contiguous lithofacies of 50%-75% sand surrounds the outer shoal and extends eastward.

The depositional-slope surface sediments are divided into the shoal-front suite, which lies immediately landward of the shoal crest, and the shoal-base suite, which lies at the landward termination of the depositional slope (Penland et al., 1986). The shoal-base suite consists of poorly sorted, very fine sand having a mean grain size of 2.9-3.3 ϕ and a standard deviation of 1.3-1.6 ϕ (Fig. 7). The shoal-front sand is coarser than the shoal-base sand and moderately well sorted; the mean grain size is 2.7-3.1 ϕ, and the standard deviation is 0.5-0.9 ϕ. The coarsest sand on the shoal surface is along the shore-parallel crest of Ship Shoal. Shoal-crest sands are moderately sorted and have a mean grain size of 2.5-2.9 ϕ and a standard deviation of 0.5-1.0 ϕ. Seaward across the upper erosional slope, grain size and sorting decrease. Across the lower erosional slope, grain size continues to decrease, and sorting begins to increase again. The surface of the upper erosional slope, or shoal ramp, consists of moderately sorted sands 2.7-3.1 ϕ in mean grain size with a standard deviation of 0.3-0.4 ϕ. Farther seaward of the shoal crest and shoal ramp, slopes decrease,

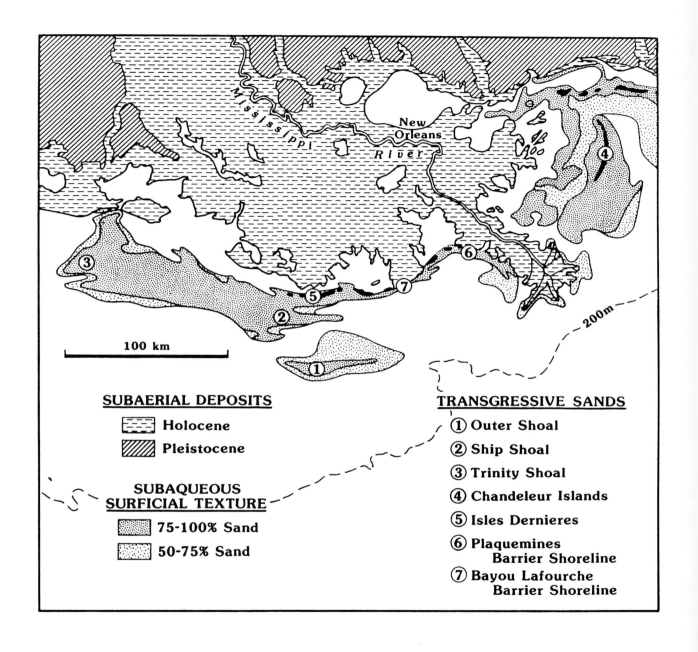

SUBAERIAL DEPOSITS

Holocene

Pleistocene

**SUBAQUEOUS
SURFICIAL TEXTURE**

75-100% Sand

50-75% Sand

TRANSGRESSIVE SANDS

① Outer Shoal

② Ship Shoal

③ Trinity Shoal

④ Chandeleur Islands

⑤ Isles Dernieres

⑥ Plaquemines
Barrier Shoreline

⑦ Bayou Lafourche
Barrier Shoreline

Figure 6. Map showing the texture of surface sediments on the continental shelf lithofacies offshore of the Mississippi River
delta plain (modified from Frazier, 1974).

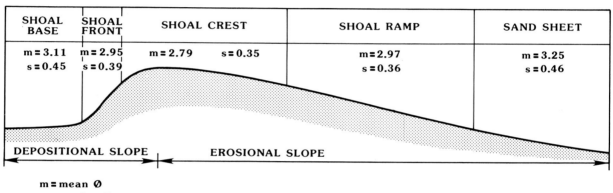

Figure 7. Diagram showing the lateral distribution of depositional environments recognized in the Ship Shoal region. Mean grain size and sorting for surface sediments in each environment are shown in phi units.

forming a zone termed the retreat path blanketed by a moderately well-sorted sand sheet that has a mean grain size of 3.0-3.4 ϕ with a standard deviation of 1.7-2.3 ϕ.

Shoal Facies

Vibracore and seismic analyses led to the identification of eight major sedimentary facies in the Ship Shoal subsurface on the basis of lithology, texture, sedimentary structures, faunal assemblages, and stratigraphic position. Major facies and sequences defined include *shoal-crest, shoal-front, shoal-base, sand-sheet*, and *lagoonal* facies overlying a regressive sequence of *distributary, delta-front*, and *pro-delta* facies. Figure 8 illustrates a composite sedimentary sequence from Ship Shoal. The shoal-crest, shoal-front, shoal-base, and sand-sheet deposits represent the stratigraphic signature of Ship Shoal as it migrates landward on the inner shelf. Table 1 lists the sedimentological parameters of these deposits. Lagoonal, distributary, delta-front, and prodelta facies are associated with deposition in paleoenvironments during stages in the development of the deltaic section underlying Ship Shoal. This manuscript will concentrate on the marine facies of Ship Shoal, the *shoal-crest, shoal-front, shoal-base*, and *sand-sheet*.

Shoal Crest

The shoal-crest facies is an accumulation of sand and shell derived from the erosion and reworking of the shoal ramp and inner shelf. The crest of Ship Shoal has the highest relative energy levels because its relief above the surrounding shelf extends into the zone of active fair-weather and storm wave processes. Water depths range from 2.7 m in the west to 7.0 m in the east. Seven vibracores penetrated the

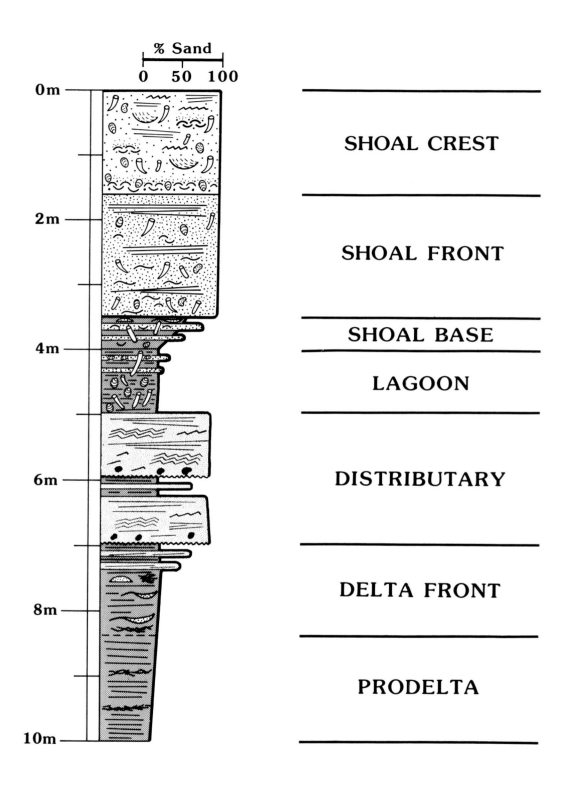

Figure 8. Composite sedimentary sequence through the entire regressive-transgressive Ship Shoal sediment package. Sedimentary facies are labeled on the right.

Table 1. Sedimentological characteristics of cored Ship Shoal facies.

Facies	Lithology and Texture	Physical Structures	Biogenic Structures
Shoal Crest	Medium sand (2.5–2.9 ϕ), moderately sorted; abundant interclasts of beach rock, Crassostrea sp. shell, and Rangia sp. shell; 95%–100% sand.	Graded (storm) bedding; low-angle planar and horizontal laminations; wave ripple laminations; medium-scale trough cross-beds.	Ophiomorpha burrows common.
Shoal Front	Fine sand (2.7–3.1 ϕ), moderately well-sorted; abundant interclasts of beach rock, Crassostrea sp. shell, and Rangia sp. shell; 90%–100% sand.	Massive appearing; low-angle planar and horizontal laminations; rare flaser bedding.	Extensively burrowed; Ophiomorpha and Skolithos burrows abundant.
Shoal Base	Very fine sand (2.9–3.3 ϕ), poorly sorted; rare interclasts of beach rock, Crassostrea sp. shell, and Rangia sp.; 65%–70% sand.	Interbeds of silty sand and silty clay; wavy bedding grades into mud flasers upwards.	Extensive burrows in silty clay beds and at tops of silty sand beds. Ophiomorpha and Skolithos burrows common.
Sand Sheet	Fine sand (3.0–3.4 ϕ), poorly sorted; rare interclasts of beach rock, Crassostrea sp. shell, and Rangia sp.; 80%–85% sand.	Graded (storm) layers; low-angle planar laminations; mud flasers.	Abundant Ophiomorpha burrows.

shoal-crest facies along the entire length of Ship Shoal, SS-1, SS-2, SS-5, SS-6, SS-12, SS-15, and SS-18. The shoal-crest facies is from 1.0-3.1 m thick. This facies consists of a clean, moderately sorted, medium-grained sand; mean grain size is 2.5-2.9 ϕ, with a standard deviation of 0.5-1.0 ϕ (Table 1). The shoal-crest deposit is 95%-100% sand. Figure 9 illustrates the shoal-crest facies with a representative grain-size histogram.

The shoal-crest facies coarsens upward and has faint horizontal and subhorizontal laminations and minor amounts of burrowing, most commonly of Ophiomorpha. Graded layers of sand and shell 10-50 cm thick are common in the shoal-crest facies and are believed to be deposited by storm processes (Fig. 10). These layers are normally graded and have sharp or erosional basal contacts; the layers are part of a bed that grades upward from sand and reworked shells, to medium-grained, parallel-laminated sands, and then to medium-scale trough cross-beds followed by symmetrical ripple laminations. Symmetrical ripple laminations reflect oscillatory wave deposition within the shoal-crest facies. Assemblages of whole and reworked modern marine mollusks, such as *Mulinia* sp. and *Olivella* sp., commonly occur. Reworked *Rangia* sp. shells, *Crassostrea* sp. shells, and lithoclasts of beachrock are incorporated in the shoal-crest facies (Fig. 10).

Shoal Front

The shoal-front facies reflects deposition on the upper depositional slope between the 4-6 m isobath at the western end of the shoal and the 8-10 m isobath at the east. It is a lower-energy environment than the shoal crest. Sediment derives from the erosion of the shoal crest and ramp. More prevalent low-energy conditions in the shoal-front

SHOAL CREST

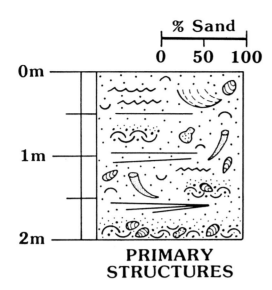

% Sand

PRIMARY STRUCTURES

- Graded (Storm) Beds
- Wave Ripple Laminations
- Low-Angle Planar Laminations
- Medium-Scale Trough-Cross Beds
- Ophiomorpha Burrows
- Clasts of Beach Rock, _Crassostrea_ sp. Shell, _Rangia_ sp. Shell

GRAIN-SIZE HISTOGRAM

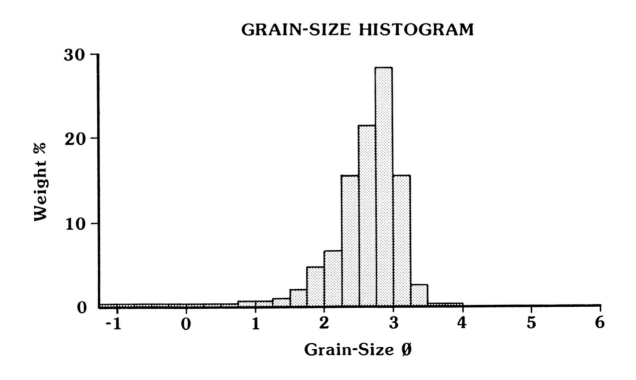

Figure 9. Diagram illustrating the primary sedimentary structures and a grain-size histogram for the *shoal-crest* facies.

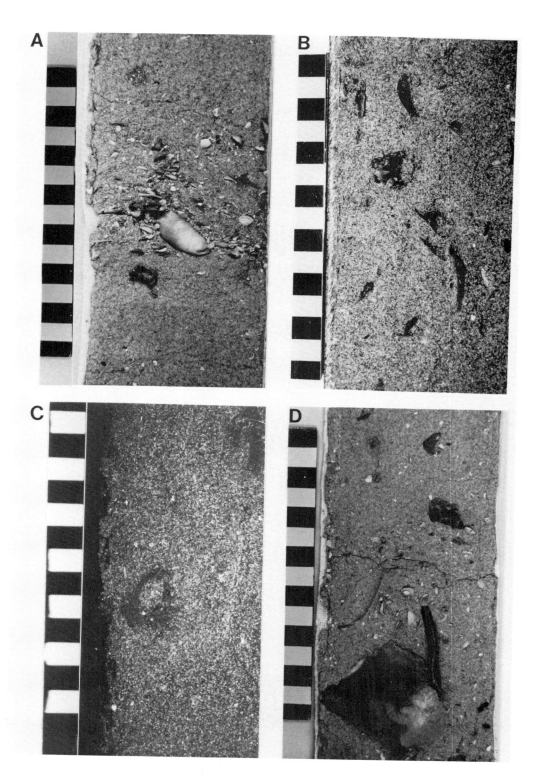

Figure 10. Representative photographs of the *shoal-crest* facies: [A] normally graded bed of *Mulinia* sp. and *Olivella* sp. shells and sand, [B] *Crassostrea* sp. shell interclasts in massive-appearing sands, [C] Ophiomorpha burrow in massive-appearing sand, [D] graded bed with beachrock clasts and *Crassostrea* sp. shell fragments (scale is in cm).

environment allow more burrowing activity than in the shallower waters at the shoal crest. Located at or below fair-weather wave base, the shoal front is not significantly reworked by fair-weather wave processes. Seven vibracores, SS-1, SS-2, SS-5, SS-6, SS-12, SS-15, and SS-18, penetrated the shoal-front facies (Fig. 2). Figure 11 illustrates the shoal-front facies and presents a representative grain-size histogram. The shoal-front facies is 1.0-3.3 m thick and consists of moderately sorted, fine-grained sand. Mean grain size varies from 2.7 to 3.1 ϕ, with a standard deviation of 0.5-0.9 ϕ. The shoal front deposit is 90%-100% sand (Table 1).

The contact between the shoal-front and the overlying shoal-crest facies is sharp; the basal contact with the shoal-base facies is gradational (Fig. 12). The shoal-front facies appears massive but is extensively burrowed, and some horizontal and low-angle laminations occur. Burrowing and shell content decrease upward in the lower-shoal sequence. The principal burrowing type is Ophiomorpha; whole and reworked shells from modern marine assemblages include *Mulinia* sp. and *Olivella* sp. shells. Fragments of *Rangia* sp. shells, *Crassostrea* sp. shells, and lithoclasts of beachrock are common.

Shoal Base

The shoal-base environment represents the advancing edge of the landward-migrating Ship Shoal depositional surface, which lies between the 8-9 m isobath at the western end of the shoal and the 11-12 m isobath at the east. Because the shoal-base environment is below fair-weather wave base, more burrowing is observed in this facies. Fifteen vibracores penetrated the shoal-base facies of Ship Shoal, SS-0, SS-1, SS-2, SS-5, SS-6, SS-7, SS-11, SS-12, SS-13, SS-14, SS-15,

SHOAL FRONT

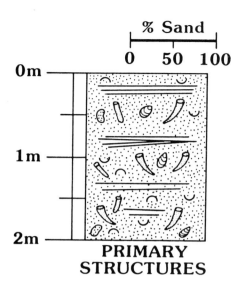

PRIMARY STRUCTURES

% Sand
0 50 100

0m
1m
2m

- Low-Angle Planar and Horizontal Laminations
- Flaser Bedding
- Ophiomorpha and Skolithos Burrows Abundant
- Clasts of Beach Rock, _Crassostrea_ sp. Shell, and _Rangia_ sp. Shell
- Massive Appearing

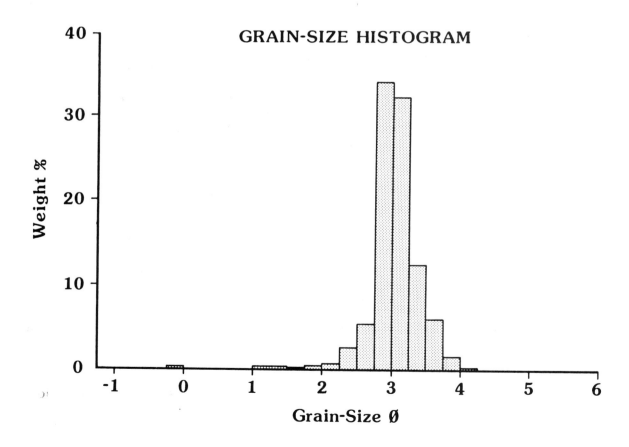

GRAIN-SIZE HISTOGRAM

Figure 11. Diagram illustrating the primary sedimentary structures and a grain-size histogram for the _shoal-front_ facies.

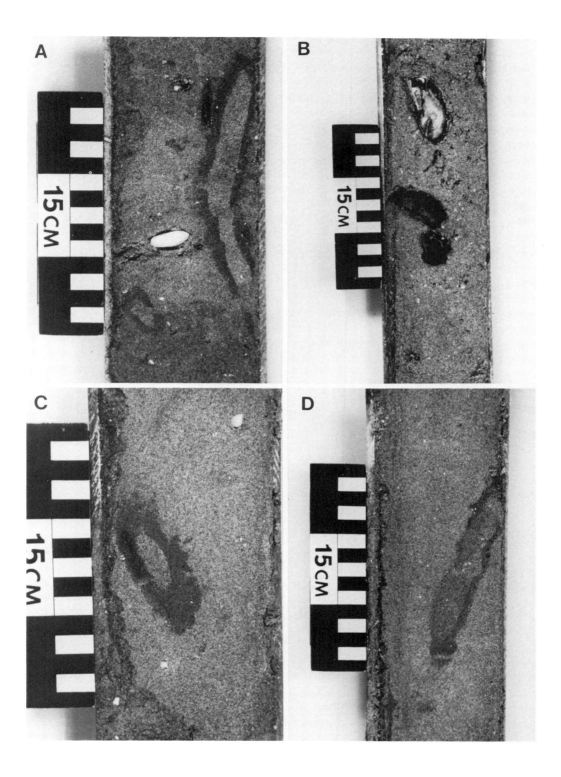

Figure 12. Representative photographs of the shoal-front facies: [A] *Ophiomorpha* burrowed sand with *Olivella* sp. shells, [B] beachrock clasts and Crassostrea sp. shell in massive-appearing sand, [C] *Ophiomorpha* burrow in massive-appearing sand, [D] *Ophiomorpha* burrow in massive-appearing sand with shell fragments (scale is in cm).

SS-16, SS-17, SS-18, and SS-19 (Fig. 2). Interbedded layers of silty clay and lenticular- to wavy-bedded, poorly sorted, very fine-grained sands characterize this facies (Fig. 13). The sand layers generally have sharp erosional bases and a gradational upper contact. The mean grain size is 2.9-3.3 ϕ, with a standard deviation of 1.3-1.6 ϕ. The shoal-base deposit is 50%-75% sand (Table 1).

The contact between the shoal base and the overlying lower shoal is gradational. Relatively extensive burrowing in the shoal-base facies increases upward. Sand and shell typically fill the Ophiomorpha and *Skolithos* burrows (Fig. 14). Silty clay layers contain wavy and lenticular beds of silt. The wavy bedding grades into lenticular bedding upward in the sequence, and the lenticular bedding increases in grain size and frequency of occurrence within the silty clay layers. Modern marine shells do occur, particularly *Mulinia* sp. and *Olivella* sp. shells, though not as commonly as in the shoal-front and shoal-crest facies. Reworked *Rangia* sp. shells, *Crassostrea* sp. shells, and lithoclasts of beachrock are rare.

Sand Sheet

A discontinuous sand sheet covers the seaward slope of Ship Shoal, marking its retreat path. The landward migration of Ship Shoal causes sediments deposited on the landward slope to be incorporated into the Ship Shoal sand body. These sediments are exhumed and re-worked again when Ship Shoal has migrated the distance equivalent to its width. At its present rate of migration (10-15 m/yr), the sand body is reworked every 500-1000 years. This pattern of shoal retreat and storm-dominated transport generates a sand sheet. Sand eroded

94

SHOAL BASE

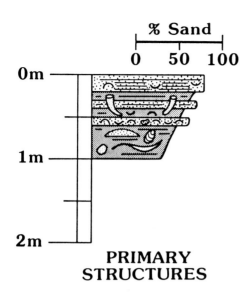

% Sand

0   50   100

0m

1m

2m

PRIMARY
STRUCTURES

- Wavy and Lenticular Bedding
- Low-Angle Planar Laminations
- Ophiomorpha and Skolithos
  Burrows Common
- Clasts of Beach Rock,
  <u>Crassostrea</u> sp. Shell
  and <u>Rangia</u> sp. Shell

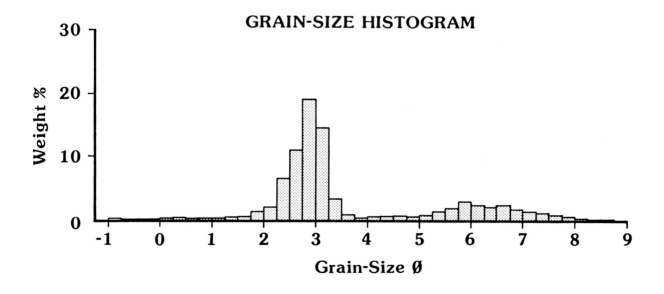

GRAIN-SIZE HISTOGRAM

Weight %

Grain-Size Ø

Figure 13. Diagram illustrating the primary sedimentary structures
and a grain-size histogram for the *shoal-base* facies.

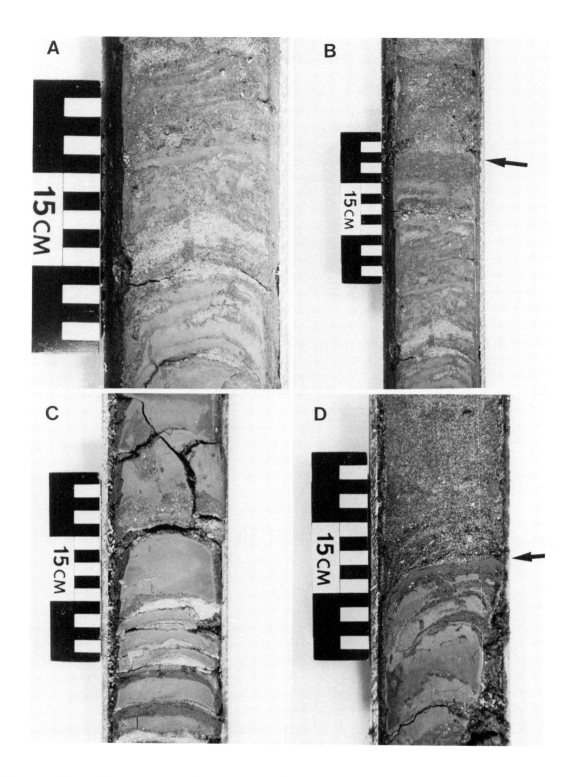

Figure 14. Representative photographs of the shoal-base facies: [A] lenticular- and wavy-bedded sands and laminated muds, [B] contact (arrow) between shoal-front (above) and shoal-base (below) facies, [C] lenticular-bedded sands interbedded with laminated muds and shell fragments, [D] contact (arrow) between shoal-base (above) and lagoonal (below) facies (scale is in cm).

from the seaward slope of Ship Shoal during storm impact is redeposited offshore, blanketing the retreat path.

Four vibracores, SS-3, SS-4, SS-16, and SS-19, cored the sand sheet and document thicknesses of 10-90 cm (Fig. 2). The sand-sheet facies is a fine sand that ranges between 3.0 ϕ and 3.4 ϕ, with a standard deviation of 1.7-2.3 ϕ (Table 1). The sand sheet consists of occasional graded bedding and flasers (Fig. 15). Horizontal and subhorizontal laminations were observed, and burrowing, primarily of Ophiomorpha, is found in the sand sheet. The basal contact of the sand sheet is erosional. Figure 16 illustrates the sedimentary features of the sand-sheet facies with representative photographs.

Lagoon

Found only in the subsurface, the lagoonal facies represents sedimentation within a backbarrier lagoon similar to the modern-day Terrebonne Bay or Barataria Bay. In this backbarrier setting, brackish-water bays evolve into saline lagoons through the processes of land loss and saltwater intrusion. Brackish-water organisms, such as the *Rangia* sp., have built many reefs in the upper ends of interdistributary bays. *Crassostrea* sp. reefs are distributed throughout the more saline lower ends of the bay behind the barrier shorelines. Ultimately, backbarrier land loss enlarges brackish bays, and shell reefs are destroyed by salt water and higher-energy wave conditions. Fine-grained sediments accumulate, and the considerable biologic activity in these backbarrier areas produces a bioturbated mud sequence. Hurricane impacts rework surface sediments and shell reefs during storm-surge inundation.

Six vibracores, SS-0, SS-1, SS-2, SS-3, SS-6, and SS-7, penetrated the lagoonal facies under the western portion of Ship Shoal (Fig.

SAND SHEET

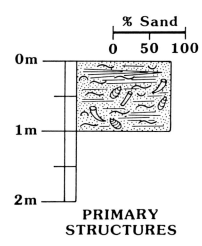

% Sand

0 50 100

0m

1m

2m

**PRIMARY
STRUCTURES**

- Graded (Storm) Beds
- Ripple Laminations
- Low-Angle Planar Laminations
- Abundant Ophiomorpha
  and Skolithos Burrows
- _Crassostrea_ sp. Shell
  and _Rangia_ sp. Shell

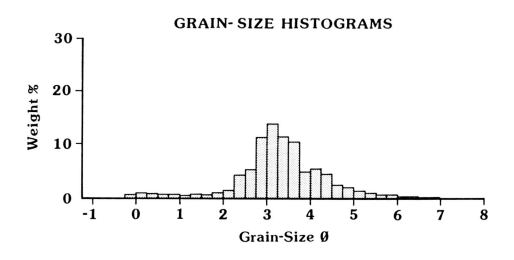

GRAIN-SIZE HISTOGRAMS

Figure 15. Diagram illustrating the primary sedimentary structures and a grain-size histogram for the *sand-sheet* facies.

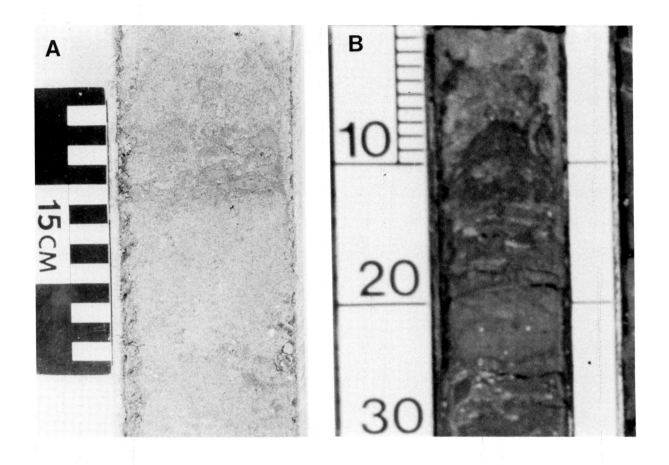

Figure 16. Representative photographs of the *sand-sheet* facies: [A] laminated sands with flaser bedding, [B] Ophiomorpha burrowed and laminated sands (scale is in cm).

2). The lagoonal facies is a poorly sorted, medium-grained silt from 0.5 to 1.5 m thick. The mean grain size ranges from 4.4 to 5.5 ϕ, with a standard deviation of 1.7-2.3 ϕ (Table 1). Sediments are between 5% and 25% sand. The lagoonal deposits are a burrowed sequence of silty clay containing parallel laminations, starved ripples, asymmetrical ripple laminations, and shell (Fig. 17). Figure 18 illustrates representative features of the lagoonal facies. Lagoonal deposits contain vertical and horizontal burrows commonly filled with sand and shells. The contact of the lagoonal deposits with the overlying shoal-base facies is a sharp disconformity. The basal contact of the lagoonal sequence with the underlying deposits is gradational. In two vibracores, thin lag deposits consisting primarily of reworked *Crassostrea* sp. shells occur at the base of the sequence.

Sedimentary Sequences

The sequence variability of Ship Shoal can best be illustrated by the vibracore logs from cores SS-6, SS-1, SS-12, SS-15, and SS-18 taken west to east along the shoal crest (Fig. 2). Inspection of these logs reveals that the vertical sequence of transgressive shoal facies is relatively uniform, in contrast to that of the underlying regressive sequence of delta facies, which displays more variability. The sequence variability of Ship Shoal is best exemplified in three regions, west (SS-6, SS-1), central (SS-12, SS-15), and east (SS-18).

The western region represents the area of maximum shoal relief and landward migration (Fig. 19). Here, the shoal sand sequence is greater than 5 m thick, and the shoal-crest deposits reach their maximum thickness of 2-3 m. The lower shoal-front deposits are 1-2 m thick and underlain by shoal-base deposits greater than 1 m thick.

LAGOON

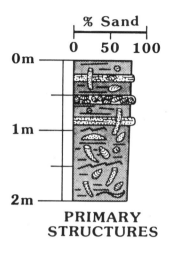

% Sand

0   50   100

0m

1m

2m

PRIMARY
STRUCTURES

- Graded (Storm) Beds
- Low-Angle Planar Laminations
- Flaser Bedding
- Clasts of Beach Rock,
  Crassostrea sp. Shell,
  and _Rangia_ sp. Shell

GRAIN-SIZE HISTOGRAM

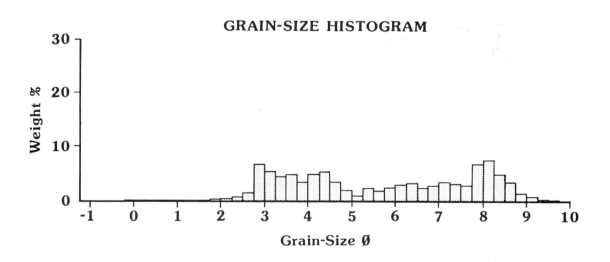

Figure 17. Diagram illustrating the primary sedimentary structures and a grain-size histogram for the _lagoonal_ facies.

Figure 18. Representative photographs of the *lagoonal* facies: [A] laminated muds with sand- and shell-filled burrows, [B] large and small burrows filled with sand in massive muds, [C] bioturbated mud, sand, and shell, [D] laminated mud having horizontal and vertical sand-filled burrows (scale is in cm).

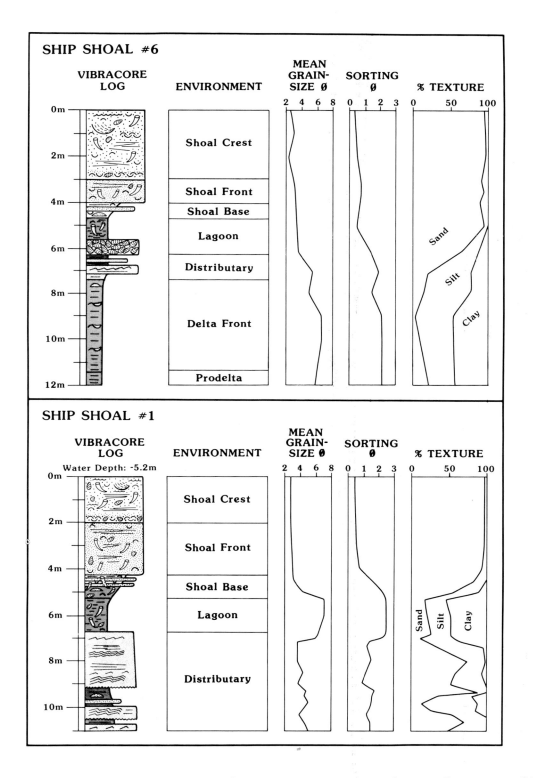

Figure 19. Representative sedimentary sequences from vibracores SS-1 and SS-6 through the crest of the western Ship Shoal region. (See Fig. 2 for location.)

The overall sequence coarsens upward. The entire shoal package lies on a sequence of lagoonal muds 1.0-1.5 m thick associated with a former backbarrier environment. In the underlying deltaic section, distributary channels are concentrated in the western shoal region near vibracore SS-1. The distributary deposits are over 5 m thick. The stratigraphic sequence observed in vibracore SS-6, taken immediately west of vibracore SS-1, defines the western margin of the subsurface distributary system. Figure 20 illustrates a photograph of Ship Shoal vibracore SS-1.

A sedimentary sequence from the central shoal region is represented by vibracore SS-12 (Fig. 21). The shoal-crest deposits are thinner than those in the western shoal region; the average thickness is 1.5 m. The shoal-front deposits are thickest in the central region and typically average 3-4 m. The shoal-base deposits are 0.5-1.0 m thick. The sedimentary sequence in vibracore SS-12 represents the eastern margin of the underlying deltaic section in the western shoal region defined by the presence of interbedded delta-front sand, silt, and clay. Farther east, the regressive deltaic sequence is composed entirely of prodelta muds.

The eastern shoal sequences are represented by vibracores SS-15 and SS-18 (Fig. 22). The thinnest shoal-crest and shoal-front facies are found in this region (1.0-1.5 m and 1.5-3.0 m, respectively). The shoal-base deposits are 1.0 m thick. The underlying regressive deltaic sequence is composed entirely of prodelta muds, which indicates sedimentation at the extreme flank of the delta system. Figure 23 is a photograph of Ship Shoal vibracore SS-15.

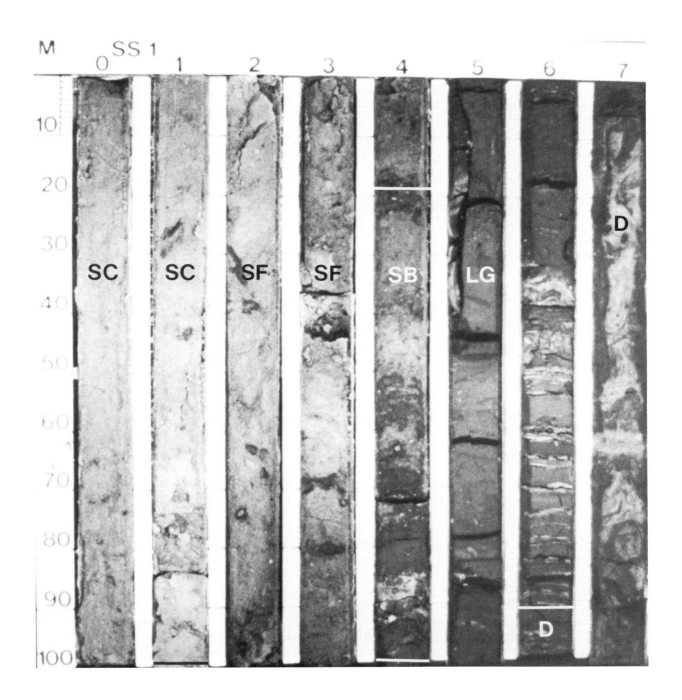

Figure 20. Photograph of Ship Shoal vibracore SS-1 from the western shoal region. The individual facies illustrated include: SC = shoal crest, SF = shoal front, SB = shoal base, LG = lagoon, and D = distributary (scale on left is in cm). (See Fig. 2 for location.)

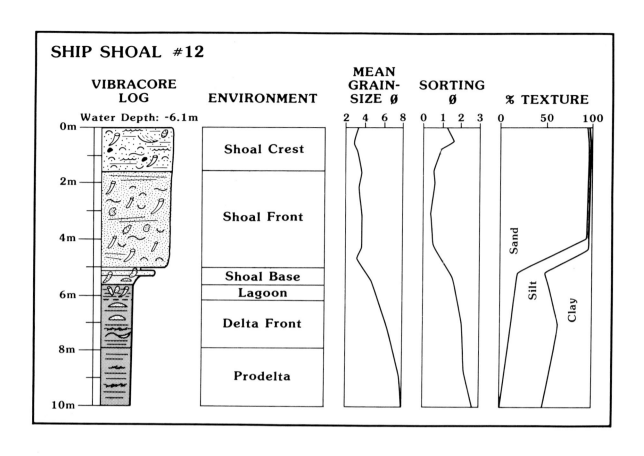

Figure 21. Representative stratigraphic sequence from vibracore SS-12 through the crest of the central Ship Shoal region. (See Fig. 2 for location.)

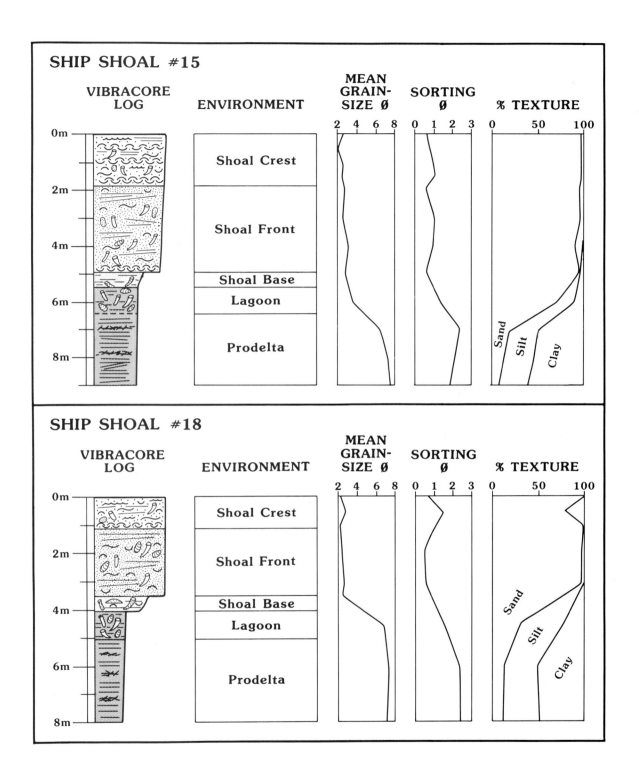

Figure 22. Representative stratigraphic sequences from vibracores SS-15 and SS-18 through the crest of the eastern Ship Shoal region. (See Fig. 2 for location.)

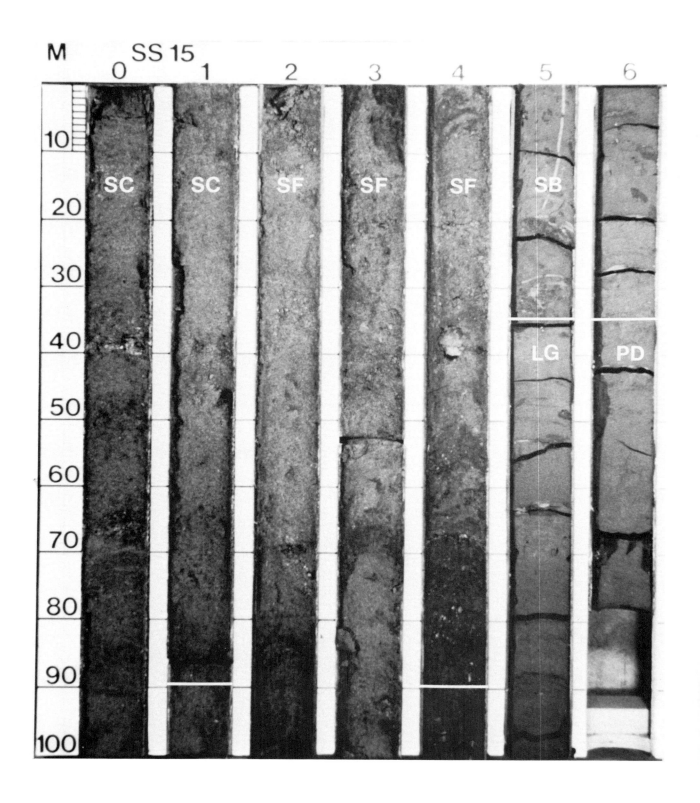

Figure 23. Photograph of Ship Shoal vibracore SS-15 from the eastern shoal region. The individual facies illustrated include: SC = shoal crest, SF = shoal front, SB = shoal base, LG = lagoon, and PD = prodelta (scale on left is in cm). (See Fig. 2 for location.)

Stratigraphic Cross Sections

Three strike-oriented stratigraphic cross sections seaward, along the crest, and landward of Ship Shoal depict the facies variability in the subsurface (Fig. 24). North to south, these cross sections are identified as (1) Ship Shoal channel, (2) Ship Shoal crest, and (3) Ship Shoal shelf. The transgressive shoal facies disconformably overlie a regressive sequence of deltaic deposits 10-12 m thick. In the western shoal region between vibracores SS-6 and SS-15, a sequence of distributary deposits 5-8 m thick extends seaward under Ship Shoal. The axis of this distributary is centered along vibracores SS-0, SS-1, SS-2, and SS-3 (Figs. 2, 24). Underlying the distributary facies is a sequence of delta-front and prodelta deposits. This regressive deltaic sequence rests disconformably on a lagoonal surface associated with an older transgressed delta. The top of this regressive sequence lies 12-14 m below sea level, and the base lies 22-24 m in the subsurface.

Overlying the regressive deltaic sequence is a blanket of lagoonal muds found throughout the Ship Shoal region (Fig. 24). The lagoonal facies is stratigraphically continuous with the underlying deltaic sequence and represents sedimentation within a backbarrier environment undergoing subsidence. The lagoonal deposits are disconformably overlain by, and are not synchronous with, the transgressive shoal sequence. Lagoonal muds 1-3 m thick are thickest over the main distributaries in the western shoal region; eastward the lagoonal muds pinch out on the erosional inner shelf.

The central strike section illustrates the stratigraphic relationships between Ship Shoal and the surrounding inner shelf (Fig. 24). West to east, the Ship Shoal sand body comprises a sequence of shoal-crest,

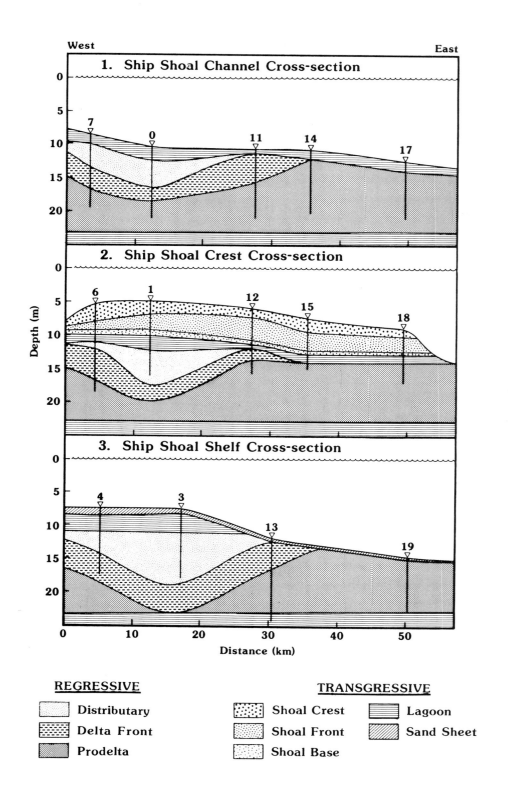

Figure 24. Stratigraphic strike sections through the Ship Shoal region. (See Fig. 2 for location.)

shoal-front, and shoal-base deposits averaging 4-5 m thick. The shoal-crest deposits thicken westward from 1.7 m to 3.5 m; concurrently, the shoal-front deposits thicken eastward from 1.2 m to 2.5 m. The under-lying shoal-base deposits average a uniform 1 m in thickness throughout the shoal region. The basal contact of Ship Shoal with the underlying lagoonal muds is sharp, and shoal-base sands lie on a disconformable lagoonal surface. The seaward cross section shows a transgressive sand sheet 10-90 cm thick spreading south from Ship Shoal onto the inner shelf.

Five dip-oriented cross sections illustrate the along-shelf stratigraphic variability in the Ship Shoal region (Fig. 25). West to east, the cross sections are identified as (A) west end, (B) west re-gion, (C) central region, (D) east region, and (E) east end. The top of the regressive deltaic sequence, upon which Ship Shoal rests, dips gently seaward and lies 10-12 m below mean sea level. Distributary channel deposits are found at the top of the deltaic sequence at the west end of the shoal (Fig. 25). The distributary sequence, 4-6 m thick, quickly thins east and west away from the channel axis. In both strike and dip sections, the transgressive shoal facies show minimal sequence variability.

At the updip (north) portion of the two easternmost cross sections, barrier island and deltaic deposits overlie lagoonal muds (Fig. 25). These facies relationships are a product of sedimentation at the distal margin of the Lafourche delta complex (Fig. 3).

Inner-Shelf Shoal Development

Much controversy exists concerning the exact nature of sand-ridge formation on the North American continental shelf. Hypotheses range

111

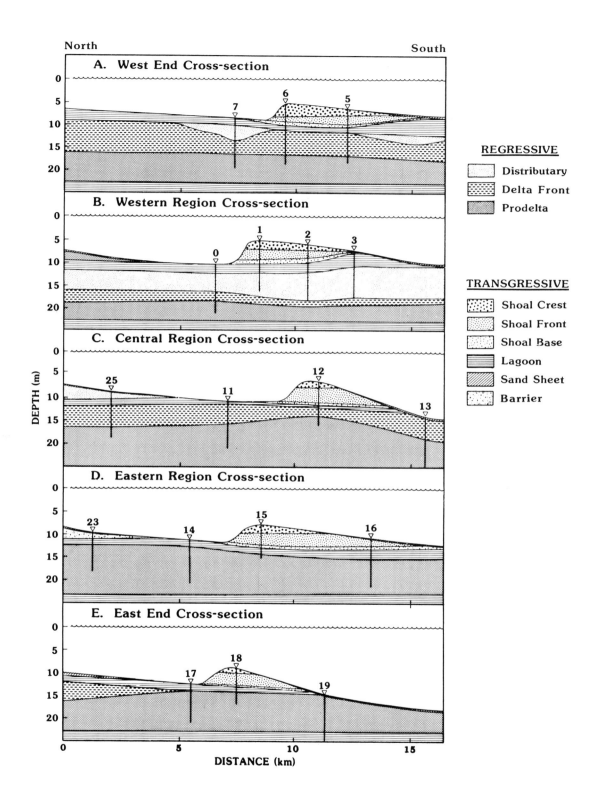

Figure 25. Stratigraphic dip sections through the Ship Shoal region. (See Fig. 2 for location.)

from those attributing sand-ridge formation to in-situ drowning of barrier shorelines by sea level rise to those attributing formation to storm-generated geostrophic flow across the shoreface and inner shelf (Rampino and Sanders, 1980; Swift and Moslow, 1982; Swift et al., 1984). The stratigraphy of Ship Shoal indicates that it represents a marine sand body originating from the transgression of a submerged barrier shoreline associated with a Mississippi River delta.

The 17 cored sequences document the fact that Ship Shoal and the underlying deltaic deposits represent a regressive-transgressive sediment package. The stratigraphic position of Ship Shoal identifies it as a transgressive sand body that has migrated landward to its present position under conditions of sea level rise, repeated storm impacts, and a predominantly southeasterly wave approach (Boyd and Penland, 1984). The vertical stacking of shoal-base, shoal-front, and shoal-crest facies documents the landward migration of Ship Shoal (Fig. 26). Overall, the Ship Shoal sand body consists of homogenous, clean sand that contains marine shells and reworked interclasts of beachrock, *Crassostrea* sp. shells, and *Rangia* sp. shells. No in-situ barrier shoreline deposits were found within the sand body of Ship Shoal.

In-situ barrier shoreline deposits would include vegetation horizons, *Crassostrea* sp. shell reefs, the washover-lagoon contact, seaward-dipping foreshore and antidune stratification, tidal inlet sequences, washover sequences, and lagoonal sequences. However, none of these deposits were vibracored in-situ within the sand body.

Reworked interclasts of beachrock, *Crassostrea* sp. shells, and *Rangia* sp. shells are common constituents of the vibracored transgressive sand sequences. The well-rounded, polished shell clasts indicate

113

SHIP SHOAL DEPOSITIONAL MODEL

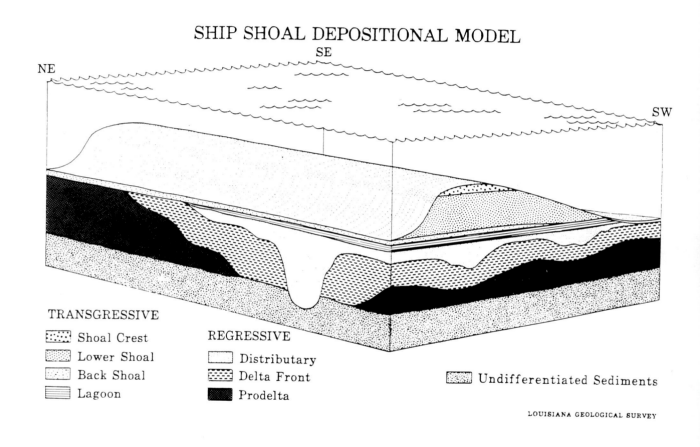

TRANSGRESSIVE

▓ Shoal Crest
▓ Lower Shoal
▓ Back Shoal
▤ Lagoon

REGRESSIVE

☐ Distributary
▥ Delta Front
■ Prodelta

▨ Undifferentiated Sediments

LOUISIANA GEOLOGICAL SURVEY

Figure 26. A block diagram illustrating the facies relationships found in the Ship Shoal region.

114

possible reworking and abrasion in a high-energy environment, such as a surf zone, during their depositional history. Beachrock, composed of sands cemented by calcium carbonate and iron, forms along all of the transgressive barrier shorelines of the Mississippi River delta plain at the water table in foreshore areas (H. H. Roberts, pers. comm.). *Crassostrea* sp. reefs commonly occur throughout the backbarrier lagoon and the lagoonal shore of flanking barrier islands and barrier island arcs. *Rangia* sp. shell reefs are common along the inland margins of transgressive backbarrier bays and lagoons. These reworked clasts are evidence of former barrier shoreline environments.

The lagoonal muds beneath Ship Shoal are the only in-situ evidence of the existence of a former barrier shoreline environment. However, these muds are not part of the transgressive inner-shelf shoal sequence. The contact between the transgressive shoal sequence and the underlying lagoonal muds is recognizable over a 5-cm, and typically no more than a 15-cm, interval. The abrupt nature of this contact suggests continuous landward shoal migration onto the inner shelf instead of sedimentation in a backbarrier lagoon. The contact generally observed between the barrier island sands and lagoonal muds in Louisiana is much thicker, as much as 1-3 m (Penland et al., 1985b).

A three-stage model from Penland and Boyd (1981) presented in Figure 27 illustrates the genesis and evolution of transgressive inner-shelf shoals. During stage 1, coastal processes transform the regressive delta into an *erosional headland with flanking barriers*. Flanking barriers are built from headland sand sources supplied by shoreface erosion through the spit-breaching process described by Gilbert (1885). During stage 2, relative sea level rise, land loss, and shoreface erosion

TRANSGRESSIVE MISSISSIPPI DELTA BARRIER MODEL

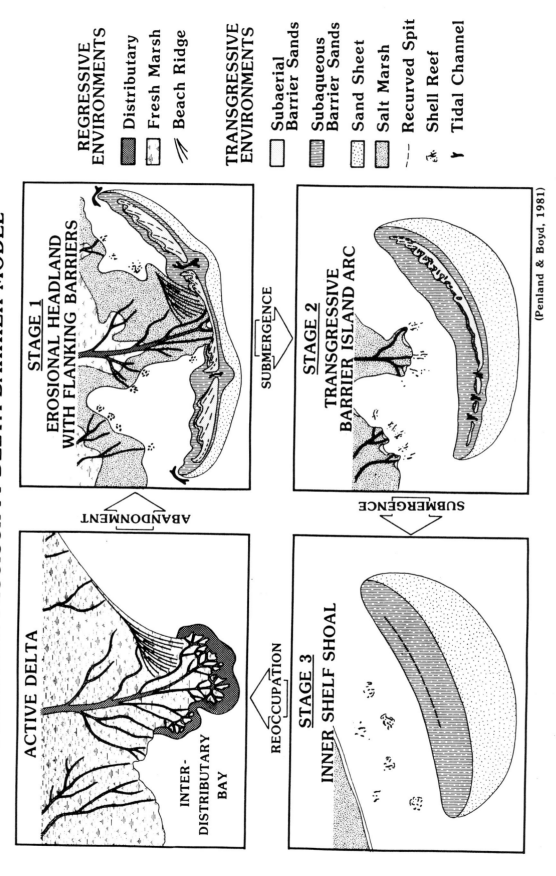

Figure 27. A three-stage transgressive barrier model proposed to explain inner-shelf shoal development offshore of the Mississippi River delta plain (modified from Penland and Boyd, 1981).

lead to the submergence and separation of the stage 1 barrier shoreline from the mainland through the detachment process described by Hoyt (1967), forming the *transgressive barrier island arc*. The final stage of barrier evolution occurs when relative sea level rise and overwash processes overcome the ability of the barrier island arc to maintain its subaerial integrity. During stage 3, submergence of the barrier island arc eventually occurs, and an *inner-shelf shoal* forms. Following submergence, the former barrier island arc sand body continues to be reworked on the shoreface and inner continental shelf.

The stratigraphic signature of *erosional shoreface retreat* is a sand sheet and ridge field (Swift and Moslow, 1982) and of *in-place drowning* an in-situ, overstepped barrier island (Rampino and Sanders, 1980). Neither model adequately explains the morphology, stratigraphy, or development of the Mississippi Delta inner-shelf shoals. The stratigraphy indicates that Ship Shoal is a marine sand shoal formed by the transgression and submergence of a Mississippi River delta barrier shoreline. The stratigraphic signature of Ship Shoal illustrates a process whereby submerged barrier shoreline sands are incorporated into the shelf record in the form of marine sand shoals, a process we term *transgressive submergence* (Penland et al., in press). During its depositional history, Ship Shoal has migrated across the shelf, allowing shoreface erosion to totally rework the original barrier shoreline sand body into a marine sand shoal (Fig. 28).

Summary

Ship Shoal is a landward-oriented sand body 50 km long and 5-12 km wide having an inner-shelf relief of 3-6 m. Ship Shoal is

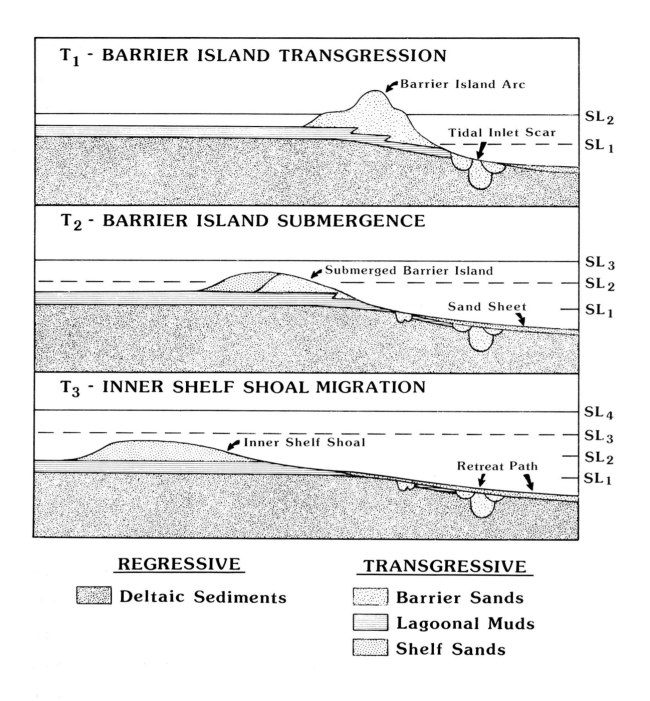

Figure 28. Transgressive submergence model explaining barrier shoreline response to relative sea level rise. The stratigraphic signature of this style of transgression is an inner-shelf marine sand body derived from the erosion of a submerged barrier island (Penland et al., in press).

comparable in size and shape to the transgressive barrier shorelines associated with the abandoned portions of the Mississippi River delta plain. The morphology of Ship Shoal can be divided into five distinct regions on the basis of surface texture, morphologic change, and shoal geometry (landward to seaward): (1) shoal base, (2) shoal front, (3) shoal crest, (4) shoal ramp, and (5) sand sheet.

Four shoal facies forming a transgressive sequence were observed from vibracore analysis of Ship Shoal: the shoal-crest, shoal-front, shoal-base, and sand-sheet. This sequence disconformably overlies a lagoonal facies that is stratigraphically conformable to the underlying regressive delta package. The current models of *erosional shoreface retreat* and *in-place drowning* do not adequately explain the geomorphology and stratigraphy of Ship Shoal. The stratigraphy of Ship Shoal illustrates a process termed *transgressive submergence* through which transgressive barrier shorelines created by the erosion of abandoned Mississippi River deltas are transformed into marine sand shoals and incorporated into the continental shelf stratigraphic record.

Acknowledgments

Support for the research results presented in this core workshop manuscript was provided by the Louisiana Geological Survey as part of the statewide Nearshore Sand Resource Inventory. Louisiana Universities Marine Consortium (LUMCON), located in Cocodrie, Louisiana, and the Sea Grant Auxiliary in Baton Rouge provided logistical field support. The authors gratefully acknowledge the field assistance of Steve Anderson, Robert S. Tye, Robert Cuomo, and Kevin Neese. Karen Westphal drafted the illustrations, Dolores Falcon typed the manuscript,

Mary Penland edited the manuscript, and Karen E. Ramsey managed the data base.

References

Boyd, R., and Penland, S., 1984, Shoreface translation and Holocene stratigraphic record in Nova Scotia, southeastern Australia, and the Mississippi River delta plain, in Greenwood, B., and Davis, Jr., R. A., eds., Hydrodynamics and sedimentation in wave-dominated coastal environments: Spec. issue of Marine Geol., v. 63, p. 391-412.

Coleman, J. M., and Gagliano, S. M., 1964, Cyclic sedimentation in the Mississippi River deltaic plain: Gulf Coast Assoc. Geol. Socs. Trans., v. 14, p. 67-82.

Cuomo, R. F., 1984, The geologic and morphologic evolution of Ship Shoal, northern Gulf of Mexico: Baton Rouge, Louisiana State University, Master's thesis, 249 p.

Fisk, H. N., 1944, Geological investigation of the alluvial valley of the lower Mississippi River: Vicksburg, Miss., U.S. Army Corps of Engineers, Waterways Experiment Sta., 78 p.

Frazier, D. E., 1967, Recent deposits of the Mississippi River, their development and chronology: Gulf Coast Assoc. Geol. Socs. Trans., v. 17, p. 287-311.

Frazier, D. E., 1974, Depositional episodes, their relationship to the Quaternary stratigraphic framework in the northwestern portion of the Gulf basin: Austin, University of Texas, Bureau of Economic Geology, Geol. Circ. 74-1, 28 p.

Gilbert, G. K., 1885, The topographic features of lake shores: U.S. Geological Survey 5th Annual Report, p. 87-88.

Gould, H. R., 1970, The Mississippi delta complex, in Morgan, J. P., ed., Deltaic sedimentation, modern and ancient: Soc. Econ. Pal. Min. Spec. Publ. 15, p. 3-30.

Hoyt, H. H., 1967, Barrier island formation: Geol. Soc. Amer. Bull., v. 78, p. 1125-35.

Kolb, C. R., and Van Lopik, J. R., 1958, Geology of the Mississippi River deltaic plain, southeast Louisiana: Vicksburg, Miss., U.S. Army Corps of Engineers, Waterways Experiment Sta., Technical Report 3-483, 120 p.

Krawiec, 1966, Recent sediments of the Louisiana inner continental shelf: Houston, Rice University, Ph.D. dissertation, 50 p.

Nummedal, D., Cuomo, R. F., and Penland, S., 1984, Shoreline evolution along the northern coast of the Gulf of Mexico: Shore and Beach, v. 52, no. 1, p. 11-17.

Penland, S., and Boyd, R., 1981, Shoreline changes on the Louisiana barrier coast: IEEE Oceans, v. 81, p. 209-19.

Penland, S., Boyd, R., Nummedal, D., and Roberts, H., 1981, Deltaic barrier development on the Louisiana coast: Trans. Gulf Coast Assoc. Geol. Socs., v. 31, p. 471-76.

Penland, S., Suter, J. R., and Moslow, T. F., 1985a, The development and stratigraphy of Ship Shoal, Northwest Gulf of Mexico, in Berryhill, Jr., H. L., Moslow, T. F., Penland, S., and Suter, J. R., Shelf and shoreline sands, northwest Gulf of Mexico: Amer. Assoc. Petrol. Geol. Continuing Education Short Course, New Orleans, Louisiana.

Penland, S., Suter, J. R., and Boyd, R., 1985b, Barrier island arcs along abandoned Mississippi River deltas, in Oertel, G. F., and Leatherman, S. P., eds., Barrier islands: Marine Geol., v. 6, p. 197-233.

Penland, S., Suter, J. R., and Moslow, T. F., 1986, The geology of the Ship Shoal inner-shelf region, northern Gulf of Mexico: Baton Rouge, Louisiana Geological Survey, Coastal Geology Bulletin Number 1, 95 p.

Penland, S., Suter, J. R., and Boyd, R., (in press), Transgressive depositional systems of the Mississippi River delta plain--a model for shoreline and shelf sand development: Jour. Sed. Petrol.

Rampino, M. R., and Sanders, J. E., 1980, Holocene transgression on south-central Long Island, New York: Jour. Sed. Petrol., v. 50, p. 1063-80.

Scruton, F. P., 1960, Delta building and the deltaic sequence, in Shepard, F. P., Phleger, F. B., and van Andel, T. H., Recent sediments, northwest Gulf of Mexico: Amer. Assoc. Petrol. Geol. Symposium Volume, p. 81-102.

Swift, D. J. P., and Moslow T. F., 1982, Holocene transgression in south central Long Island, New York, discussion: Jour. Sed. Petrol., v. 52, no. 3, p. 1014-19.

Swift, D. J. P., McKinney, T. F., and Stahl, L., 1984, Recognition of transgressive and post-transgressive sand ridges on the New Jersey continental shelf, in Tillman, R. W., and Siemers, C. T., Siliciclastic shelf sediments: Soc. Pal. Min. Spec. Publ. No. 34, p. 25-36.

SEDIMENTARY CHARACTERISTICS

OF MODERN STORM-GENERATED SEQUENCES:

NORTH INSULAR SHELF, PUERTO RICO

Robert S. Brackett[1,2] and David M. Bush[1]

[1]Department of Geology, Duke University, Durham, North Carolina

[2]Present Address: Shell Offshore Inc., New Orleans, Louisiana

Abstract

High-energy events (i.e., storms) play an important role in deposition of modern sediments on the north insular (island) shelf of Puerto Rico, a high-energy, mixed siliclastic/carbonate terrain. Individual storm sequences, such as the SS1, have been analyzed and correlated along the northwestern shelf in the area of Arecibo. The SS1 storm deposit is characterized by a distinctive sedimentary sequence representing three phases of deposition: (1) wave-generated basal lag deposit formed during "peak" storm conditions, (2) graded sand and silt unit redeposited from suspension during waning of storm energy, and (3) a poststorm flood deposited layer. The three units of this individual storm sequence were recognizable by variations in grain size, percent calcium carbonate, and primary physical structures.

A sharp, erosional contact and a well-developed lag marks the base of the SS1 sequence. The lag is much coarser grained and higher in calcium carbonate content than the underlying sediment. Both the high calcium carbonate and increased grain-size values are attributable to accumulations of shelly gravel and the selective removal of fine terrigenous sediment during peak storm winnowing. As the storm wanes, suspended sediment is deposited as a graded sand unit often containing possible hummocky cross-stratification. This portion of the SS1 sequence is characterized by decreasing-upward grain-size and carbonate values. The finest grain-size and lowest carbonate values of the SS1 deposit occur in the poststorm inundite. Between events, physical structures within the inundite and upper portion of the tempestite are obscured by bioturbation.

Faunal constituents within the SS1 basal lag change rapidly with increased distance from the Arecibo River. Faunal changes are correlatable with the change from fluvially derived (allochthonous) terrigenous sedimentation to shelf-derived (autochthonous) carbonate sedimentation. Near the river mouth (terrigenous-allochthonous sedimentation) the primary lag constituents are pelecypods and gastropods. The abundance of echinoids, bryozoans, and algal-encrusted fragments increases with the change to carbonate-autochthonous deposition away from the river mouth. A relative change in calcium carbonate content and median grain size also occurs.

The distance from the Arecibo River mouth seems to be the most important factor determining the characteristics of the SS1 event deposit. The following trends are observed in the SS1 sequence with increasing distance from the river mouth: (1) increased percentage of shelly carbonate in the basal lag deposit (from about 5% to over 70%), (2) increased thickness of the graded storm-sand unit (from 12 to 21 cm), (3) increased bioturbation levels (from rare to abundant burrows), (4) decreased preservation (as burrowing increases), (5) decreased thickness of the poststorm flood layer (from 16 to 4 cm), (6) decreased displacement of faunal constituents, (7) decreased frequency of hummocky cross-stratification and other wave-generated primary structures, and (8) decreased amalgamation of storm deposits.

Introduction

Recognition of modern storm deposits (tempestites) has increased the awareness of the tremendous influence that storms have on shelf sedimentation (Hayes, 1967; Reineck and Singh, 1972; Kumar and Sanders, 1976; Howard and Reineck, 1981). However, despite broad interest and increased study, details of shelf hydrodynamics, storm transportation processes, sediment emplacement mechanisms, and other storm-related processes remain obscure (Kreisa, 1981). Present knowledge of storms and storm deposits comes mostly from research on ancient shelf sediments. This is partially due to (1) the inherent difficulties and dangers in making observations during storms in modern settings, (2) limitations of sample size (studying cores instead of outcrops), and (3) the possibility that present rates of bioturbation are greater today than in the past (Kreisa, 1981). An increase in observed bioturbation through time implies that modern storm sequences are less likely to be preserved in the geologic record than pre-Jurassic deposits (Thayer, 1979; Brandt Velbel, 1985).

Perhaps the greatest problem in studying modern (Holocene) storm deposits is the widespread occurrence of relict sediments on most continental shelves (Kreisa, 1981; Swift et al., 1971). These nonequilibrium shelves, therefore, are commonly sand-rich (noncohesive sediments) and provide poor analogs to the most ancient mud-rich (cohesive sediments) continental shelf settings. Also, recognition of storm deposits is often difficult because evidence of the "catastrophic" origin of storm units is not always obvious (Kreisa, 1981).

The north shelf of Puerto Rico provides an excellent setting for the study of storm sedimentation. Two major rivers--the Rio Grande de Arecibo and the Rio Grande de Manati--discharge significant quantities of terrigenous sediment to the shelf study area. The majority of this sediment is delivered to the shelf only after substantial rainfall on the island's central highlands. Episodic river discharge should allow the separation of flood events through recognition of individual flood deposits (inundites).

The objectives of this study are to (1) determine the importance of storms in shelf sedimentation, (2) correlate a single storm event (SS1) across the Arecibo shelf portion of the study area, (3) establish criteria to recognize depositional phases within an individual storm unit, and (4) develop a model of storm sedimentation for a high-energy shelf.

Regional Setting

Puerto Rico is the smallest and easternmost major island of the Greater Antilles Chain (Fig. 1A). It is nearly rectangular in shape and measures approximately 160 by 50 km. Distinct differences in geology, topography, and climate allow the island to be subdivided into three physiographic provinces: (1) the Central Cordillera, (2) the north coast karst region, and (3) the north and south coastal lowlands (Western Geophysical, 1974).

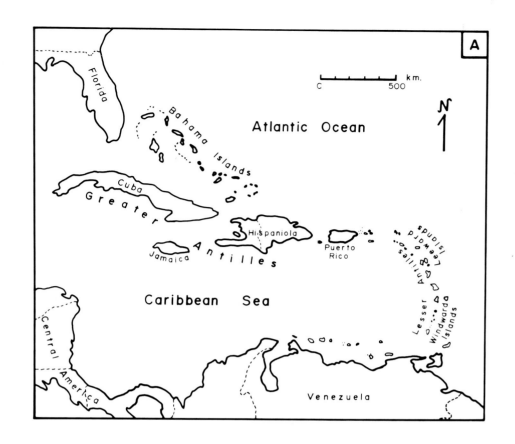

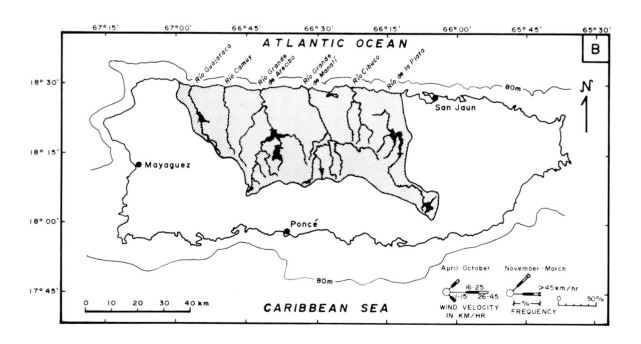

Figure 1. [A] Regional location map for Puerto Rico. [B] Map of Puerto Rico showing shelf break around island (80 m), major rivers draining into north shelf study area, and wind velocity data from San Juan. This paper deals mainly with the shelf area directly off Rio Grande de Arecibo.

The insular north shelf of Puerto Rico ranges in width from 2 to 5 km, with the shelf slope break at approximately 80 m water depth. The study area is a storm-wave dominated environment. Since Puerto Rico is located in the zone of the northeasterly trade winds, east and northeast winds predominate (see wind rose, Fig. 1B). North Atlantic winter storms and tropical disturbances can alter this picture. No comprehensive wave height or current meter data are available. However, the authors have observed that, during all seasons, seas 0.5-1.5 m are common and typically last from late morning to early evening. According to Fields and Jordan's (1972) study of storm wave swash, the Arecibo shelf has wave swash heights of 5.5 m above mean sea level at a recurrence interval of approximately six years and of +3.9 m at least once per year. Hurricanes that pass within 150 m of San Juan (63 km to the east) have a recurrence interval of less than two years (Secretary of the Army, 1962). Tidal range is low, less than 0.8 m in San Juan. Current measurements made by Wood et al. (1975a, 1975b) show the dominant current direction to be to the west (typical maximum of approximately 40 cm/sec). Although storms, which move much of the sediment, may come from any direction, Morelock et al. (1985) have observed coastal geomorphic features along the central part of the north coast of Puerto Rico and concluded that net longshore sediment transport is to the west. This, perhaps, implies that storm-driven currents are not randomly oriented here and may even have a fairly consistent trend (as proposed by D. J. P. Swift, pers. comm. to O. H. Pilkey).

The crest of the central Cordillera in Puerto Rico is located closer to the south coast. Because of the northeast trade winds, the largest rainfall is received on the northern part of the island (Lopez and

Colon-Dieppa, 1973). Thus, the greatest river discharge occurs along the north coast, and the bulk of the island's terrigenous sediment is transported by rivers to the north shelf (Ehlman, 1968; Calvesbert, 1970). That a disproportionate area of Puerto Rico drains to the north is illustrated in Figure 1B. Hurricanes and tropical storms can cause greatly increased rainfall, often resulting in severe flooding and episodic fluvial discharge events (Lopez, 1964).

Surficial sediments on the north shelf have been studied extensively. Bush (1977) expanded on original work by Schneidermann et al. (1976) by grouping shelf sediments into five distinct suites based on composition, carbonate fraction physical condition, calcium carbonate percentage and content, mean grain size, and mud content (Fig. 2). Distribution of these sediment types off the Rio de la Plata River led Pilkey et al. (1978) to conclude that the area was a site of equilibrium shelf sedimentation. That is, sediment supply, sediment cover, and physical processes are in equilibrium. Further work has allowed extrapolation of that model to the entire study area (Pilkey et al., 1984). Grossman (1978) used X-ray diffraction of heavy minerals to "fingerprint" sediments from the Manati and Arecibo rivers and showed that along-shelf transport of terrigenous sediment has formed a mixed sand belt between the two river mouths. Shelf areas adjacent to river mouths are dominated by fluvially derived, allochthonous sedimentation, whereas shelf areas removed from fluvial influence are sites of shelf-derived, autochthonous carbonate accumulation. Allochthonous deposits of modern, dark-colored, muddy, fine sands, and a minor admixture of preserved, "fresh-looking" skeletal carbonates, form subaqueous wedges immediately offshore of rivers. These wedges may have

130

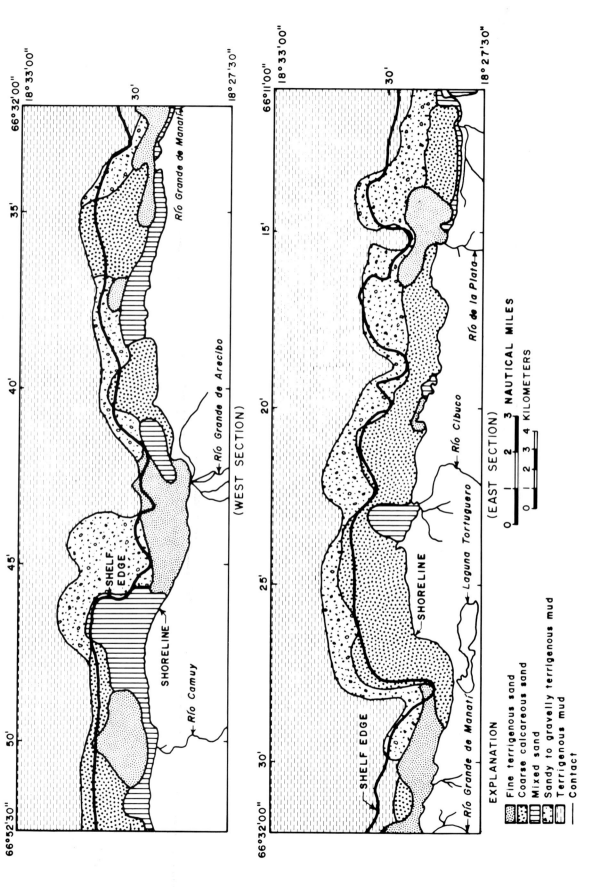

Figure 2. Bathymetry (shelf edge) and surficial sediment types of the north shelf study area (from Pilkey et al., 1984).

131

a maximum thickness of 40 m but are more likely to average around 20 m (Figs. 3, 4; Pilkey et al., 1978). In the inter-river portions of the shelf, the slowly accumulating, light-colored skeletal carbonates are quickly degraded by physical, chemical, and biological means (Pilkey et al., 1979).

Storm Sedimentation

Three different types of storms affect the north shelf of Puerto Rico. Tropical cyclones result in increased wave energy and significant river flooding. River flooding may also result from storms associated with higher-than-usual precipitation in the central highlands without an accompanying increase in wave activity. The third possibility is generation of large waves related to swells from distant, Atlantic storms associated with no rainfall or increased river discharge.

Increased wave energy from a hurricane results in formation of both tempestites and inundites. The tempestites are characterized by a sharp erosive lower boundary, a basal lag deposit, and a graded sand layer (Fig. 5). Significant river discharge from storm precipitation results in the introduction of large amounts of sediment to the shelf and the resulting deposition of an "inundite."

When significant river discharge occurs with no increase in wave energy, fine terrigenous sediment is delivered onto a quiet shelf and deposited as a thin "mud blanket." Subsequent cross-shelf transport during the next period of increased wave activity may disperse the sediment to the outer shelf and slope. Preservation of this deposit on the high-energy shelf is unlikely except in topographic depressions (Pilkey et al., 1978; Grove et al., 1982).

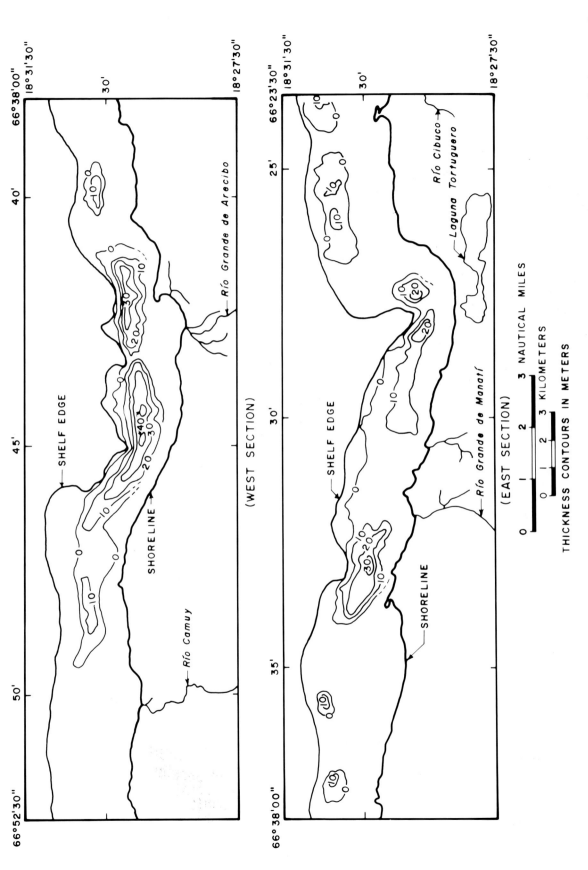

Figure 3. A plot of approximate thickness of unconsolidated sediment cover on a portion of the north shelf. Thicknesses are based on the fact that two-way travel time is approximately equal to 1 m of thickness (from Western Geophysical, 1975).

133

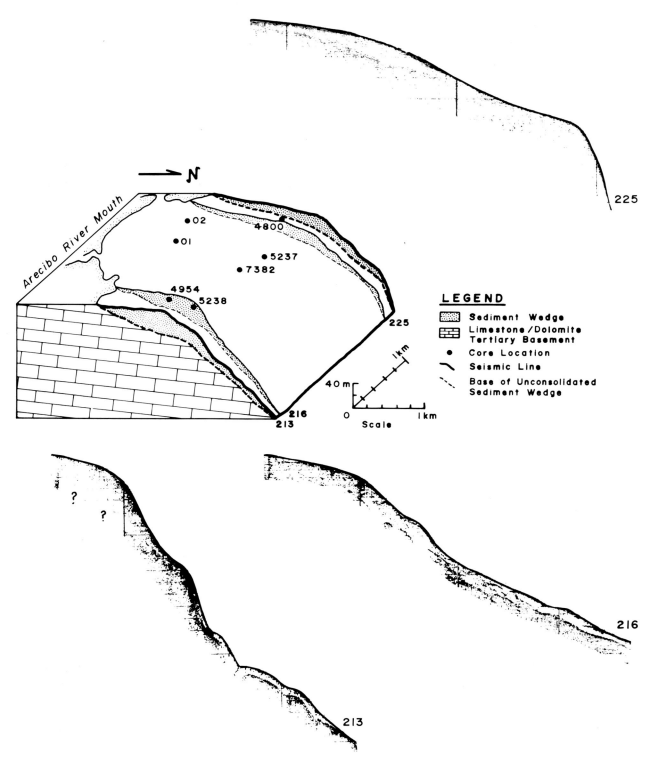

Figure 4. Schematic view of the Arecibo sediment wedge. Drawn in
the block diagram are cross-shelf "slides" that are interpretations of the four Western Geophysical high-resolution
(Uniboom) seismic lines. Uninterpreted versions of lines
213, 216, and 225 are shown for comparison.

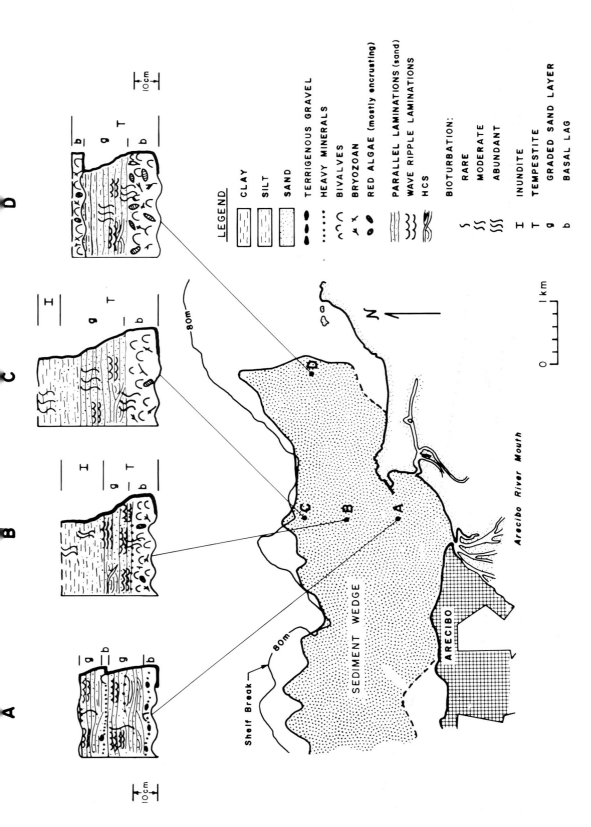

Figure 5. Generalized storm sequences from the Arecibo shelf. A, B, and C are from directly off the river mouth, in the main part of the sediment wedge. D is from east of the river mouth, on the fringe of the sediment wedge.

LEGEND

CLAY
SILT
SAND
TERRIGENOUS GRAVEL
HEAVY MINERALS
BIVALVES
BRYOZOAN
RED ALGAE (mostly encrusting)
PARALLEL LAMINATIONS (sand)
WAVE RIPPLE LAMINATIONS
HCS

BIOTURBATION:
  S   RARE
  SS   MODERATE
  SSS ABUNDANT

I   INUNDITE
T   TEMPESTITE
g   GRADED SAND LAYER
b   BASAL LAG

135

Large waves generated by storms as much as 1600 km and more from Puerto Rico have resulted in extensive beach erosion and destruction of ocean front buildings on the north coast. A storm in December 1962 remained stationary for 24 h over the North Atlantic Ocean off the coast of Nova Scotia, Canada, resulting in waves 8 m high on the Puerto Rico coast (Fields and Jordan, 1972). These storms could generate storm lags (if appropriate material is present) and suspend large quantities of sediment. The erosive capability of such large waves on the sea floor is enormous and can result in amalgamation of storm units deposited from previous storm events.

Applying Swift's (1976) model of allochthonous shelf sedimentation to the north coast of Puerto Rico would show the process by which sediments pass through the coastal zone to be river mouth bypassing. Since the shelf here is so steep (1:40, or about 1.3° compared to "typical" continental shelf values of 1:500 or 0.1°), the rate of shoreline translation during the Holocene transgression has been slow. Slow shoreline translation, according to the Swift model, allows the estuaries to fill with sediments. Thus, during floods, sediments are introduced directly onto the shoreface.

The high "normal" wave energy of the study area allows any potential reworking and redistribution of sediments to occur quickly. Thus, after flood-derived sediments reach the shelf, they are quickly (within months) dispersed, and an "equilibrium stance" is attained (Grove et al., 1982).

Many of the world's modern clastic shelf environments occur off large, sediment-trapping estuaries. Little, if any, modern, fluvially derived sediments are introduced into the shelf transport system. The results are nonequilibrium shelves with relict sediments that are probably coarser and cleaner than shelves throughout much of geologic time. This makes the northern shelf of Puerto Rico a more likely analog to ancient shelf environments and an ideal study area from which to develop a storm-sedimentation model.

Methods

Vibracores and piston cores utilized in this research were obtained between February 1984 and January 1985 aboard the R.V. Cape Hatteras and the Puerto Rico Department of Natural Resources, R.V. Jean A. The midshelf (20-40 m water depth) was sampled using an Alpine vibracore. The maximum recovery was 9.5 m of 10-cm diameter, lined core. A Ewing piston core was used on the outer shelf and on the upper slope (in water depths greater than 40 m). Inner-shelf cores were obtained in August 1984 with a diver-operated hydraulic "jackhammer" vibracore deployed from the Jean A. (see Shinn et al., 1981, for a vibracore description).

A storm-generated sedimentary deposit was recognized and correlated among 32 cores (Fig. 12), 7 of which were selected for subsampling (Figs. 4, 7). Resin peels (relief casts) were made from each core section to enhance primary sedimentary structures and aid in core description.

The resolution of depositional boundaries within the correlated storm sequence can be a function of sampling density; therefore, the selected cores were subsampled continuously at 2-cm intervals from below the storm unit base to an overlying flood deposit. This sampling pattern results in the best boundary resolution and still provides sufficient sample for analysis. One core (4800) was sampled at 10-cm intervals because no obvious storm deposit was recognizable.

The Storm Sequence Event

An individual storm sequence (SS1) has been correlated among 32 cores within the study area (Figs. 4, 13). The thickness of this storm sequence varies from 30 to 50 cm on the midshelf to 1-10 cm on the outer shelf. On the inner shelf, the sequence has been truncated by wave action but may have originally been greater than 1 m thick. The SS1 sequence is believed to have been deposited as a result of large hurricane-generated waves and poststorm flooding.

The base of the SS1 sequence is often marked by a sharp erosional contact and a well-developed lag (Fig. 5). The storm-generated lags are composed primarily of skeletal carbonate grains and are commonly overlain by parallel-laminated sand and silt. Laminations are more likely to be preserved in the lower section of the tempestite sequence since primary physical structures in the upper portions are usually obliterated by bioturbation. The upper boundary of the tempestite or inundite is difficult to recognize and could only be determined by detailed textural analysis (and even then, not with absolute confidence). The top of the sequence is typically bioturbated or truncated by subsequent wave activity.

Placing an absolute date on a storm sequence is very difficult. Storms erode, resuspend, and transport previously deposited sediments, making accurate dating impossible. For example, lags from autochthonous shelf regions may contain degraded skeletal carbonates mixed with modern, nondegraded material. The organic carbon in sediment samples from cores 5237 and 5238 was radiometrically dated using the

138

C-14 technique. The samples were treated with dilute hydrochloric acid to remove all the calcium carbonate so that only the fluvially introduced organic matter would be dated. We hoped that by dating a preserved inundite layer, the effects of resuspension by storms could be neglected, and a more representative date would be obtained.

Figure 6 schematically shows the relative position of the SS1 unit within the shelf sediment pile. The cores in Figure 6 correspond to cores A, B, and C in Figure 5. Cores 5237 and 5238 correspond to hypothetical core B in Figures 5 and 6. The dated material from core 5237 was taken from a well-preserved inundite layer 439 cm from the top of the core and 326 cm below the base of the SS1 unit and yielded an age of 6770 ± 180 C-14 years BP. The material dated from core 5238 was taken from 125 cm from the top of the core and 11 cm below the basal lag and dated at 4670 ± 80 C-14 years BP. It must be noted that since the material dated in core 5238 was from just below the SS1 unit, and could have been reworked many times before being buried, the resulting date is a maximum age of the SS1 "event." The age determination suggests that only 125 cm of sediment was deposited at this site since 4670 C-14 years BP. This is noteworthy and poses an interesting problem. The thickness of the unconsolidated sediment cover in the study area averages about 20 m (ranges up to 40 m). It presumably has all been deposited during the Holocene transgression (maximum of approximately 18,000 years). This means the minimum net accumulation rate is approximately 111 cm/1000 years. Core 5238, however, shows only 125 cm net accumulation in almost 5000 years. That is less than one-fourth the minimum average rate. It is likely

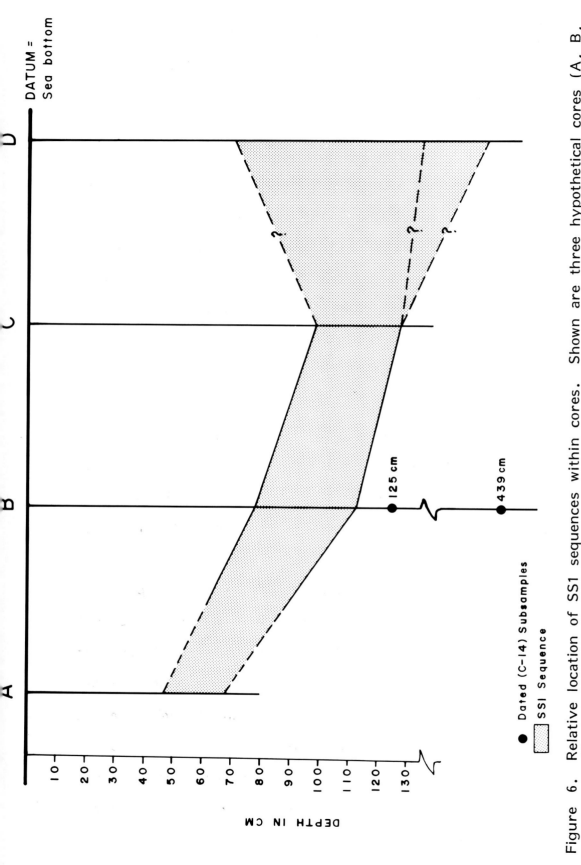

Figure 6. Relative location of SS1 sequences within cores. Shown are three hypothetical cores (A, B, C) forming the same cross-shelf transect as in Figure 5, and one core (D) from east of the river mouth, on the fringe of the sediment wedge. Dashed lines indicate where amalgamation of sequences makes precise correlation very difficult. Core B represents cores 5237 and 5238, from which were taken samples for C-14 dating (see text).

that the answer is closely tied into changes in the rate of sea level rise through the Holocene.

Presumably, wave energy is intense enough here to preclude sediment ever building up to sea level. Furthermore, a minimum sea level curve for the tropical western Atlantic by Lighty et al. (1982) suggests that the rate of sea level rise slowed considerably approximately 6000 years ago. At that time, sea level was perhaps within 6 m of its present location. A decreasing rate of sea level rise, and probably no change in sediment influx, means proportionately less and less volume (space) on the shelf in which to accumulate sediments. This means that either the shelf must "fill up" with sediments, or in this high-energy environment, more and more newly introduced sediment must bypass the shelf. The second explanation seems more logical. This suggests that, assuming sediment influx has remained constant, net sediment accumulation changes with time. Therefore, the present sedimentation rate of approximately 25 cm/1000 yr is much slower than the rate 10,000 years ago.

Sedimentary Characteristics

Basal lag. Lag deposits are formed during peak storm conditions and represent reworked material that has been winnowed in place (Kreisa and Bambach, 1982). Lags may also be formed by an interruption of sediment supply, continuous current activity, or shallowing and regression (Bloss, 1982).

In the study area, SS1 lag deposits were recognized by the sharp basal contact, abundance of skeletal carbonate, and/or increase in median grain size relative to the underlying sediment (Fig. 7). Lag

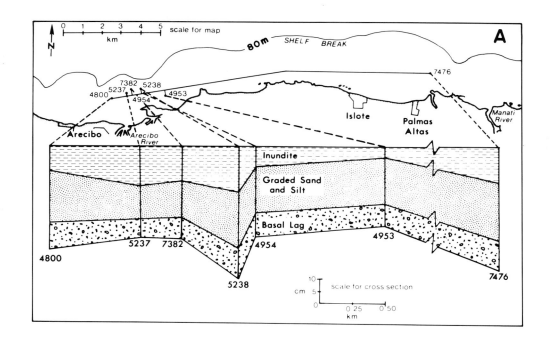

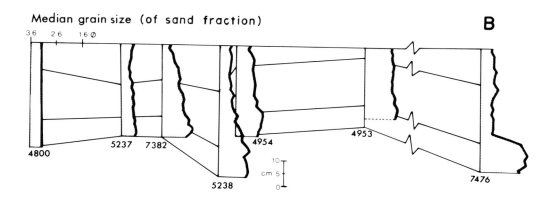

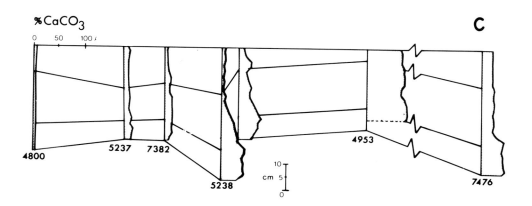

Figure 7. "Fence diagrams" of selected cores in study area to show various trends. [A] Thickness trends in SS1 unit. [B] Median grain size trends (0 units, SS1 unit). [C] CaCO₃ percentage trends in SS1 unit.

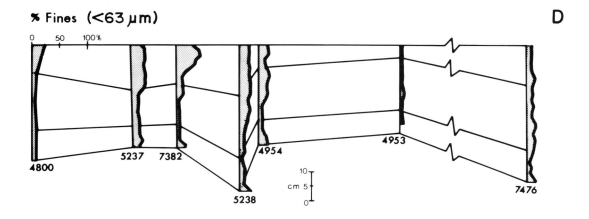

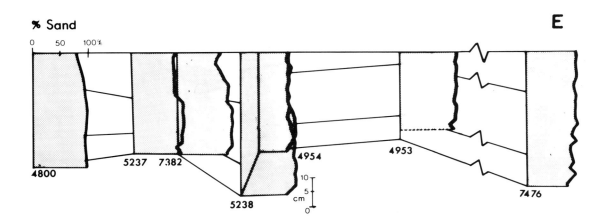

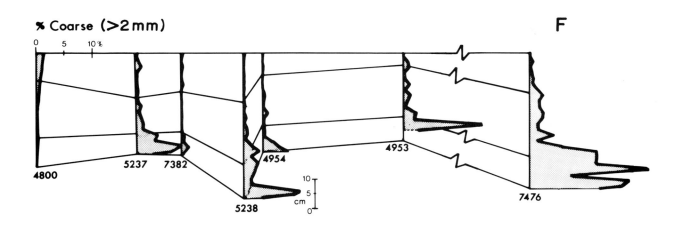

Figure 7. [D] Percentage fines (< 62 μm) in SS1 unit. [E] Percentage sand (62 μm 2 mm) in SS1 unit. Note similarity to SP or GR log response ("blocky" appearance). [F] Percentage shelly gravel (> 2 mm) in SS1 unit.

thickness of the SS1 sequence varies from 7 to 12 cm in the cores sampled and averages 9.4 cm, or approximately 25% of the total storm sequence. In areas of the shelf where autochthonous carbonate sediments are accumulating (Pilkey et al., 1978), the lags are composed of an admixture of fresh and degraded (older) carbonate material. Most of the carbonate material in the lags is degraded because it (1) accumulates so slowly and (2) degrades so quickly (Fig. 8A; Pilkey et al., 1979). Closer to river mouths, in the allochthonous-terrigenous shelf areas, the concentration of degraded carbonate decreases and the percentage of rock fragments and fresh skeletal grains increases (Fig. 8B).

Whole and fragmented pelecypods, arthropods, and igneous and metamorphic rock fragments are the most abundant constituents of the SS1 sequence in the muddy sediment adjacent to the Arecibo River (Fig. 9). East of the river, the percentages of bryozoans, foraminifera, red algae, and algal-encrusted fragments all increase. The presence of in situ faunal assemblages and the fresh and angular condition of many shells within the lag deposit suggests that significant transport of the coarser skeletal lag components on the midshelf was unlikely during generation of the SS1 storm lag.

In many modern shelf environments, comparative analyses of life and death assemblages from depths below the normal wave base have shown little, if any, between-habitat mixing of skeletal components (MacDonald, 1976; Warme et al., 1976). However, mixing of faunal assemblages in nearshore, beach, and intertidal areas is typical of most shelves (Kreisa, 1981). Aigner and Reineck (1982) concluded that mixed faunas and allochthonous elements are characteristic of proximal

144

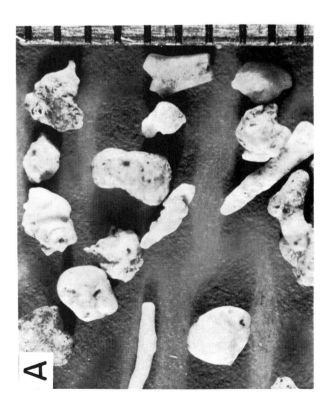

Figure 8. Photomicrographs of the very coarse sand fractions of the two "end-member" sediment types. [A] From carbonate-autochthonous (shelf-derived) sediment, typical degraded skeletal material that is "old-looking." [B] From terrigenous-allochthonous (river-derived) sediment with typical fresh carbonate fraction.

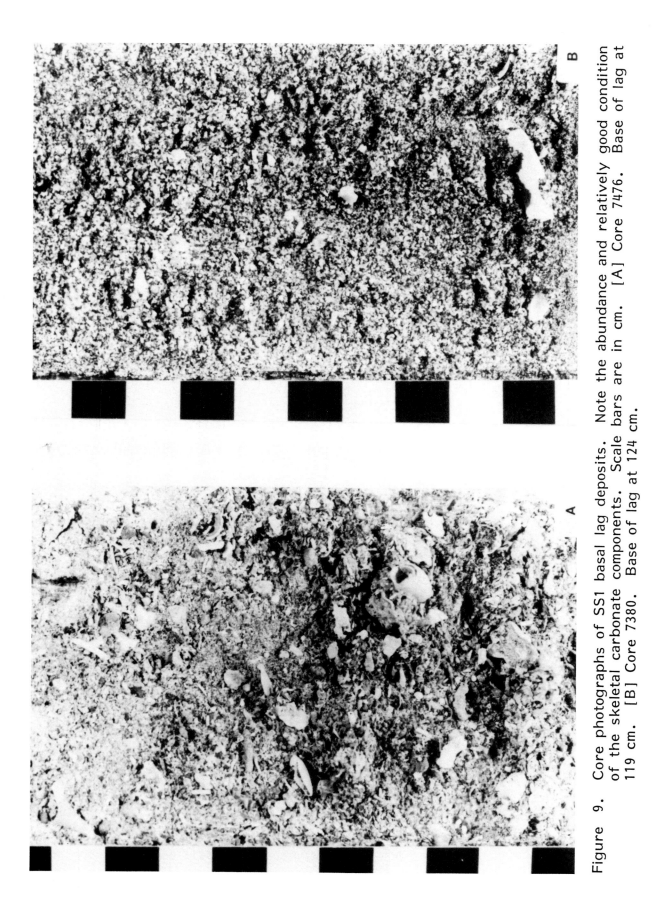

Figure 9. Core photographs of SS1 basal lag deposits. Note the abundance and relatively good condition of the skeletal carbonate components. Scale bars are in cm. [A] Core 7380. Base of lag at 119 cm. [B] Core 7476. Base of lag at 124 cm.

146

shell-rich tempestites, whereas distal tempestites are dominated by winnowed, autochthonous, soft-bottom assemblages.

A well-developed basal lag is not present in every cored SS1 sequence observed. In order to develop a lag deposit, suitable material must exist for separation during high-energy winnowing. The very well sorted sediments off the Arecibo River (e.g., core 4800) contain only small concentrations of shelly gravel (> 2-mm-size fraction) and thus are not suitable for lag development (Fig. 7F). In areas where heavy minerals are more abundant (e.g., west of the Arecibo River), the base of a storm deposit may instead be marked by a concentration of reworked heavy minerals. The thickest SS1 lag deposits were observed east of the Arecibo River, where the sediment is most enriched with coarse carbonate grains (Figs. 7A, 7C). Large storms often "cannibalize" previously deposited storm sediments and concentrate the coarse material into one lag deposit, erasing all evidence of previous weaker storms (Seilacher, 1982). A subsequent storm, unless it is more intense, will not have coarse material available for lag generation. Accordingly, when two large storms of equal intensity strike the Puerto Rico coast, the resulting sediment deposit, and especially the lag of the first storm, will appear much more "catastrophic" than that of the second storm.

Graded storm sand layers. As the peak of the SS1-generating storm passed, basal lags (where present) were rapidly overlain by a graded layer of sand and silt. Sediment thrown into suspension during peak storm conditions is quickly redeposited when oscillatory currents induced by storm waves are no longer capable of keeping the sediment in suspension. In a study off Fire Island, New York, Kumar and

Sanders (1976) estimated that storm waves are capable of scouring the nearshore sand bottom to depths as great as 2 m. All of this sediment may go into turbulent suspension and be distributed throughout the water column. As storm energy begins to wane, the suspended sediment is rapidly deposited as a normally graded layer. Rapid deposition of storm sand layers is evidenced by occasionally preserved biogenic escape structures.

Graded storm sand layers of the SS1 sequence may display parallel and wave-ripple laminations, but primary sedimentary structures are often disrupted by bioturbation (Fig. 10). Laminations are very well preserved in nearshore cores where wave energy prevails over biologic activity. As water depths increase offshore, waves "feel" the bottom less frequently. As a result, there is an offshore increase in biologic activity and therefore an offshore decrease in the preservation of primary physical sedimentary structures.

Laminated storm sand layers can form by deposition from suspension clouds in slowly moving water and have been simulated experimentally in the laboratory (Reineck and Singh, 1972). Kreisa and Bambach (1982) attribute parallel laminations in Paleozoic shelf sands to deposition from storm-induced suspension clouds. Most likely, storm sand layers in the SS1 sequence were also deposited from wave-generated suspension clouds. Storm sand layers within the SS1 sequence tend to be graded, moderately to well sorted, and exhibit a decreasing upward calcium carbonate percentage (Fig. 7C). The graded sand units in the SS1 sequence vary between 12 and 22 cm thick (Fig. 7A) and make up an average of 46% of the total storm sequence thickness. The proportion of storm sand layer thickness to total storm

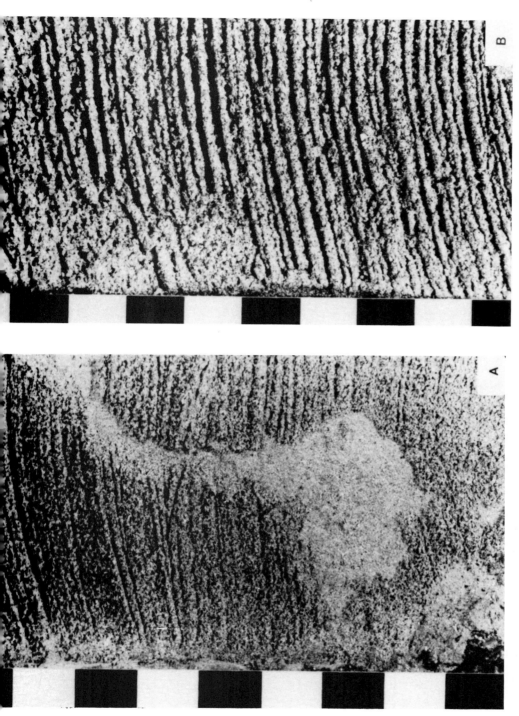

Figure 10. [A] Well-preserved burrow and laminations from a nearshore core (core 01) in 5 m water depth from directly off the Arecibo River. Disruption and reworking of primary physical sedimentary structures increases in an offshore direction. Note low-angle truncations (possible hummocky cross-stratification) in top one-third of photo. Scale bars are in cm. [B] Well-preserved parallel laminations from a nearshore core (07) in 7 m water depth from directly off the Arecibo River. Disturbance of lamina in upper left of photo is attributed to coring operation. Scale bars are in cm.

149

unit thickness is probably related to regional grain size differences. Regions with coarser-grained sediment (autochthonous) contain proportionately thicker storm sand layers than those in relatively fine-grained, muddy areas proximal to the river mouths.

A storm-generated sequence in the nearshore cores (Fig. 12) provides the best example within the study area of what may be hummocky cross-stratification. The sequence starts with a basal lag 4 cm thick containing small pebbles and coarse sand, which is overlain by approximately 10 cm of parallel-laminated fine sand having low-angle truncations. The top of the sequence contains ripple-laminated fine sediment with abundant organic detritus.

The upper portions of hummocky cross-stratified sequences in nearshore cores are often truncated. For example, core 01 contains a parallel-laminated sand sequence with an erosional truncation and deposition of another sequence by a later storm event (Fig. 12C).

Poststorm flood layer. The final phase of deposition of the SS1 storm sequence is a flood layer or inundite. An inundite capped storm deposit is only possible when intense poststorm flooding occurs. Rivers reach peak discharge as long as two days after a hurricane has passed; therefore, most of the river sediment is discharged onto the shelf after the seas have returned to near-normal conditions. It is possible that a muddy layer could result from resuspension and redeposition of shelf muds and thus be a part of the tempestite. However, in the case of the SS1, we believe the muddy cap to be a true inundite. The combination of increasing percent $CaCO_3$ away from the river (Fig. 7C), the thinning of the unit away from the river mouth (Fig. 7A), and the sharp erosional contact (not gradational) shown in Fig. 11B all point to

150

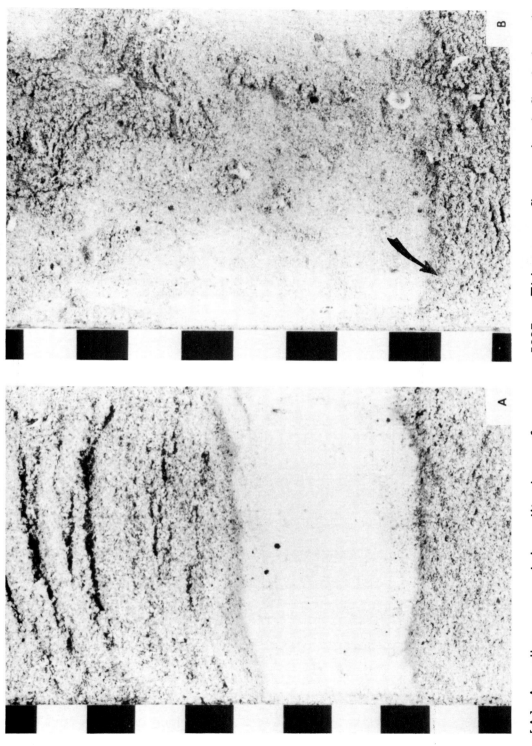

Figure 11. [A] A well-preserved inundite layer from core 5237. This very fine-grained flood deposit showing almost no relief in the photo is not part of the SS1 event. It is pictured here to illustrate the sharpness of the upper and lower contacts of the inundite when well preserved. Scale bars are in cm. [B] Photograph of the SS1 tempestite-inundite boundary. The contact is near the bottom of the photograph (see arrow) and is marked by the change from graded sand and silt to terrigenous mud (core 5237). Scale bars are in cm.

Figure 12. Possible examples of hummocky cross-stratification. Scale bars are in cm. [A] Parallel and wavy laminations with low-angle truncations (core 01). [B] Note change from parallel to wavy laminations and low-angle truncations (core 02). [C] Example of erosional truncation and amalgamation of a storm sequence by a subsequent storm (core 0.1).

Arecibo River flooding as the source, and thus the term "inundite" is properly applied.

The thickness of the SS1 poststorm flood layer is a function of distance from the Arecibo River. Close to the river, poststorm flood layers comprise the bulk of the total storm deposit thickness, while in more distal areas the flood layer thins (Fig. 7A). For example, the SS1 flood deposit in core 5237, located directly off the Arecibo River, is 16 cm thick and comprises 46% of the total storm sequence in that core. Core 4953, which was taken from a site 2 km east of core 5237, contains an inundite deposit only 4 cm thick, or 13% of the SS1 sequence thickness (Fig. 7A).

The tempestite-inundite contact is often diffuse from bioturbation and only recognizable by detailed textural analysis. X-radiography may have helped in this regard but was not utilized in this study. When the overlying inundite deposit is thick, the probability that the contact will be preserved increases. That is the case with core 5237, which has a 16-cm-thick flood deposit and a well-preserved contact (Fig. 11A).

The inundite portion of the SS1 sequence differs from the underlying tempestite in several ways. Poststorm inundites are very well sorted and characterized by a relatively low calcium carbonate content, decreased grain size, increased percentage of silt and clay, and absence of shelly material. The sand content is low and may remain constant throughout the layer (Fig. 7).

The upper inundite boundary is marked by an increase in grain size and calcium carbonate content. The original upper boundary is usually altered because of amalgamation by subsequent storm events, fair-weather winnowing, or biologic activity (Fig. 11B). Storm and flood events increase the concentration of available nutrients, creating an environment favorable to burrowing organisms.

After hurricanes David and Frederick in 1979, surficial sediment sampling of the shelf revealed a thin but widespread mud layer. Repeated sampling six months later showed that the majority of the mud had been removed by subsequent wave activity. Cross-shelf transport and removal of storm-flood layers by subsequent storms has been called "mud-hopping" (Pilkey et al., 1978). Drake et al. (1972) documented similar cross-shelf transport of a flood deposit following the major 1969 flood of the Santa Clara and Ventura rivers in Southern California. However, "mud-hopping" may not always effectively transport all the sediment introduced to the shelf during large flood events, particularly if events are frequent. Therefore, cores may contain significant amounts of silt- and clay-size material in inundite layers. During large floods more fine-grained sediment is deposited than can be removed; therefore, some record of the event may be preserved. Preservation potential increases if another flood event occurs soon enough to bury the previous deposit.

Discussion

Bioturbation and Preservation Potential

The preservation potential of the SS1 sequence, and of storm-event deposits in general, is a function of water depth, storm event frequency (amount of time for bioturbation), and storm intensity. An increase in the amount of biogenic structures with increasing water depth decreases the preservation potential of storm units offshore.

The thickness of a storm sequence is a function of storm intensity, and thick storm deposits are often well preserved. The considerable thickness of the SS1 deposit (up to 51 cm) has enabled good preservation of the tempestite portion of the sequence.

On the outer shelf (> 50 m water depth), a decrease in thickness of event deposits and an increase in the effects of bioturbation make preservation (and recognition) difficult. However, increased accumulation rates favor preservation of rare, very high energy deposits. Increased flood frequency increases the likelihood of preservation of previous deposits by burying them deeply and quickly.

Nearshore cores in the study area contain well-preserved primary structures and little evidence of disruption by biologic activity. These cores contain stacked storm sequences, each of which truncates the underlying sequence. However, on the midshelf, bioturbated units separate the storm sequences. In these cores, remnant parallel laminations are common, but laminations are frequently altered by bioturbation (Fig. 5).

Shelf depositional settings are commonly classified according to the relative degree of bioturbation. Howard and Reineck (1981) compared depositional facies of the high-energy California shelf (Ventura-Port Hueneme area) to an offshore sequence on a low-energy beach at Sapelo Island, Georgia. They noted a seaward increase in the effects of bioturbation at both localities but observed that the depositional controls on preservation of primary sedimentary structures varied. They concluded that the relative water depth at which preservation of primary sedimentary structures decreases is a function of wave energy. Nearshore, transition, and offshore facies were defined for the two study areas based on the relative dominance of physical (wave-formed) structures versus biogenic structures (Howard and Reineck, 1981). Accurate assignment of a nearshore-transition facies boundary for Puerto Rico shelf sediments is not possible because the lack of cores from the inner shelf between 8 and 20 m (see Fig. 6) prohibits facies definition. This boundary may be designated as the greatest depth at which the "fair-weather" wave base transports the bottom sediments. The presence of storm units in cores at depths greater than 23 m suggests that the "transition zone" on the Puerto Rico shelf extends to a greater depth than that of the California shelf (19 m).

Calcium Carbonate Content

The calcium carbonate content of shelf sediments within the SS1 sequence increases rapidly with distance from the Arecibo and Manati rivers (Fig. 7C). This is due to increased terrigenous influx in the interriver areas. Average calcium carbonate values within the SS1 sequence vary from 6% in core 4800 (located off the Arecibo River) to 65% in core 4953 (located east of the river). The carbonate content is as high as 76% in the storm-wave generated shell lag of the SS1 sequence in core 4953 (Fig. 7C).

The relative abundance of calcium carbonate is one of the most useful criteria for recognition of storm deposit boundaries. A sharp increase in carbonate content relative to the underlying sediment marks the base of each SS1 sequence. This increase results from in situ winnowing during peak storm conditions. Rapid deposition of suspended sediment during periods of waning storm energy is marked by a decreasing-upward calcium carbonate content from basal lag values, whereas the overlying poststorm inundite layer commonly contains a relatively lower carbonate content. The top of an event is marked by return of the calcium carbonate values to equilibrium values.

The changes observed in the calcium carbonate content within each SS1 storm sequence are attributable to a direct relationship between the grain size and calcium carbonate content of the sediments studied. The silt-and-clay-size fractions of the sediments is composed mostly of terrigenous material and contains less than 10% calcium carbonate. The coarse-grained sediment fraction is mostly skeletal carbonate material. Therefore, when grain size increases, the calcium carbonate content also increases (compare Fig. 7B to Fig. 7C). In core 5238, the amount of calcium carbonate in the sediment drops from 45% in the basal lag to 10% at the top of the graded sand-and-silt unit. The inundite unit is characterized by sediment which contains less than 10% carbonate (Fig. 7C).

Grain Size and Composition

Sand-size fraction. The median of the sand-size (0.0626-2.0 mm) fraction increases laterally away from the Arecibo River because of the increase in abundance of skeletal carbonate grains. Down-core grain size variation is minimal in the terrigenous-allochthonous areas close to the Arecibo River, where every sample is very well sorted and exhibits a median in the very fine sand range. The median grain size grades laterally to a fine sand in cores east of the Arecibo River and decreases offshore (Fig. 7B).

The effect of storms on the median grain size distribution is much more pronounced in areas of the shelf where the sediment is not very well sorted. During peak storm energy (lag development) fine sediment is winnowed away, resulting in a sharp increase in grain size at the SS1 base. As the storm wanes, a graded sequence of sediment is deposited out of suspension. In very fine grained sediment, the storm response is minimized, and no grading or lag development is observed (e.g., core 4800, Fig. 7B).

Primary terrigenous constituents in the sand-size fraction are quartz, volcanic rock fragments, magnetite, and hornblende. Primary skeletal components are molluscs, bryozoans, red algal fragments, Halimeda plates, foraminifera, echinoid spines, and sponge spicules.

Silt-and-clay-size fraction. Sediment within the SS1 sequence usually contains less than 15% silt-and-clay-size material (<0.0625 mm), except for the poststorm inundite unit, which contains a greater concentration (Fig. 7D). The basal lag of the SS1 sequence occasionally contains a significant amount (< 10%) of fine sediment (e.g., core 5238). This mud in the high-energy lag may originate from mud trapped inside shells but could also have infiltrated during waning storm conditions. In Paleozoic shell lags, Kreisa and Bambach (1982) have proposed that abundant infiltration fabrics, such as shelter porosity, sediment screening, and micrograded sediment on individual shells, occur as a storm passes and the finer sediment is deposited from suspension.

The silt-and-clay-size sediment within the SS1 sequence is primarily fluvial mud. This is confirmed by the low calcium carbonate content. Carbonate grains generally make up less than 10% of this size fraction. Heavy minerals (magnetite, hornblende, and epidote) and quartz are common constituents.

Shelly-gravel-size fraction. The "shelly-gravel"-size fraction (>2 mm) of sediments in the SS1 sequence is seldom greater than 1%, except in carbonate-autochthonous areas and basal lags. Such lags may be composed of up to 20% shelly gravel (Fig. 7F).

The concentration of shelly gravel decreases in an upward direction in the graded storm sand unit, then decreases to a minimal value (< 1%) and remains fairly constant in the overlying fluvially dominated poststorm inundite deposit (Fig. 7F).

The primary constituents of the shelly-gravel-size fraction in the SS1 sequence are skeletal carbonate grains. In very nearshore cores, volcanic and metamorphic gravels are common. Due to the increased skeletal carbonate content of shelf sediments away from the river mouths, the percentage of shelly gravel in SS1 shelf sediments increases (except for very nearshore cores) away from the rivers.

Sediment Accumulation

The base of a correlatable storm deposit represents the same time horizon everywhere it is present and is therefore valuable for determining net sediment accumulation rates. The relative down-core depth to the base of the SS1 sequence increases offshore and toward the submarine canyon heads (Fig. 13). This trend suggests that cross-shelf and lateral sediment transport are active processes and that net accumulation rates on the outer shelf are greater than in nearshore areas. Increased sediment thickness at canyon heads indicates that a significant amount of sediment is transported to the canyons during storm events. The areas where the sediment thickness over the SS1 sequence is greatest corresponds to the thickest sediment pods observed on the regional seismic profiles (Fig. 4).

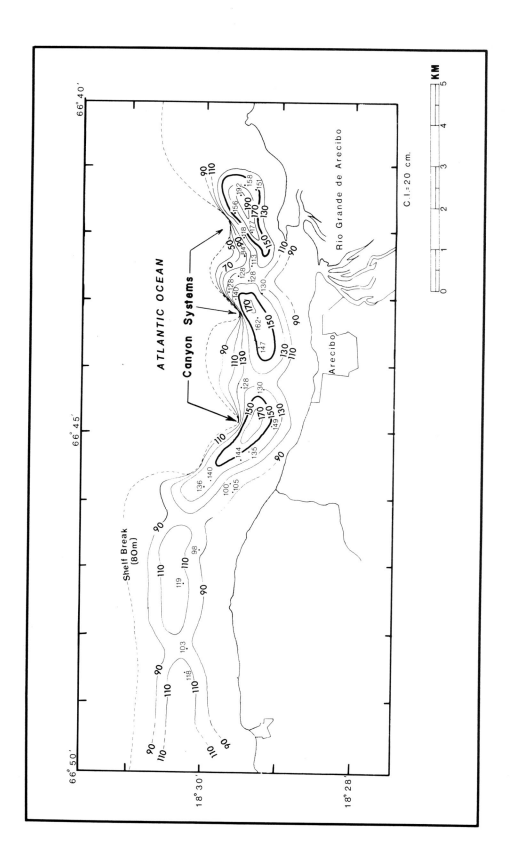

Figure 13. Isopach map of the sediment thickness (cm) above the base of the correlated SS1 unit. Note distinct along-shelf linearity and continuity.

160

Applications: Future Research

Modern/Ancient Analogs

We hope that this study has provided a modern analog for many ancient storm-dominated shelves. Examples of modern shelf study areas for comparison with ancient shelf sequences are lacking because the majority of modern continental shelves are "drowned" and out of equilibrium with modern-day processes. The abundance of relict material combined with entrapment of fine sediments in estuarine environments prohibits sediments from attaining an equilibrium stance on many modern continental shelves.

The storm response of muddy and sandy sediments can be substantially different (Kreisa, 1981). The abundance of mud in ancient open-shelf environments and lack of mud in many modern shelves do not allow application of most modern storm sedimentation models to ancient open-shelf settings. Thus, the muddy equilibrium north shelf of Puerto Rico is a perfect natural laboratory for continued research of storm sedimentation principles. Expansion of the study area to the whole north shelf, combined with detailed sampling, long-term current measurements, and extensive pre- and poststorm sampling should clear the obscurity of many storm-related processes and associated sedimentary structures. Data from shelf sediment tracer studies using tagged sand (e.g., Lavelle et al., 1978) would greatly increase our knowledge of the hydrodynamic environment on the north shelf during storm events.

Exploration Models and Potential Reservoir Quality

Paleogeographic reconstructions are a major goal of any research on ancient sediments. Proximality trends (e.g., Aigner and Reineck,

1982) may indicate storm and source, basin geometry, and paleo-bathymetry. Recognition of the relationship between water depth and storm deposition on modern shelves is essential for paleobathymetric reconstruction of ancient sequences. Expansion of lateral proximality trends to the vertical dimension could identify transgressive/regressive fluctuations, which may give insight to hydrocarbon explorationists. Thick and laterally persistent amalgamated storm sand layers interbedded with shelf muds are potential hydrocarbon reservoirs (Aigner, 1985). Large amounts of muddy fine sand can quickly be winnowed clean on the high-energy north shelf of Puerto Rico, even by fair-weather processes (Grove et al., 1982). This could leave great thicknesses of clean, good-quality reservoir sand. There is no reason to doubt that the entire preserved sediment package could be composed of clean, permeable sands (Figs. 11, 13). Additionally, the clean nearshore sands quickly grade into muddier outer-shelf and upper-slope deposits that could act as a source for hydrocarbons. Thus, thick accumulations of sediments on a storm-dominated shelf could be an excellent target for petroleum exploration. Amalgamation of storm units on modern shelves concentrates heavy minerals, which may also prove to be economically significant. Determination of the sediment thickness over a correlatable tempestite deposit (Fig. 16) should aid in mapping the locations of the thickest sand accumulations for exploration.

Acknowledgments

This work was supported by the National Science Foundation (Grant #333-0721 to O. H. Pilkey) and the U.S. Geological Survey. We wish to thank the Captain and crew of R.V. Cape Hatteras and R.V. Jean A. The Puerto Rico Department of Natural Resources assisted in the field and in the laboratory. Special thanks to Tonya Clayton and Lynne Claflin for last-minute drafting and Fran Woods for typing a few drafts of this manuscript. The authors would like to acknowledge the invaluable contributions of Rafael W. Rodriguez and Orrin H. Pilkey to this study.

References

Aigner, T., 1985, Storm depositional systems, in Friedman, G. M., Neugebauer, H. J., and Seilacher, A., eds., Lecture notes in earth sciences, v. 3: Springer-Verlag, New York, 174 p.

Aigner, T., and Reineck, H. E., 1982. Proximality trends in modern storm sands from the Helgoland Bight (North Sea) and their implications for basin analysis: Senckenbergiana maritima, v. 14 (5/6), p. 183-215.

Bloss, G., 1982, Shell beds in the lower Lias of South Germany, facies and origin, in Einsele, G., and Seilacher, A., eds., Cyclic and event stratification: Springer-Verlag, New York, p. 223-239.

Brandt Velbel, D., 1985, Preservation of "event beds" through time: AAPG Bulletin Abstracts, v. 69, No. 2, p. 240.

Bush, D. M., 1977, Equilibrium sedimentation: North insular shelf of Puerto Rico: Duke University Department of Geology, Master's thesis.

Calvesbert, R. G., 1970, Climate of Puerto Rico and U.S. Virgin Islands: Climatology of the U.S., no. 60-52, U.S. Department of Commerce.

Drake, D. E., Kolpack, R. L., and Fischer, P. J., 1972, Sediment transport on the Santa Barbara-Oxnard Shelf, Santa Barbara Channel, California, in Stanley, D. F., and Swift, D. J. P., eds., Marine sediment transport and environmental management: John Wiley, New York, p. 307-332.

Ehlman, A. J., 1968, Clay mineralogy of weather products and of river sediments, Puerto Rico: Jour. Sed. Petrology, v. 38, p. 885-889.

Fields, F. K., and Jordan, D. G., 1972, Storm-wave swash along the north coast of Puerto Rico: U.S. Geol. Survey, Hydrologic Investigations: Atlas H.A. 432.

Grossman, Z. N., 1978, Distribution and dispersal of Manati River sediments, Puerto Rico north insular shelf: Duke University Department of Geology, Master's thesis, 72 pp.

Grove, K. A., Pilkey, O. H., and Trumbull, J. V. A., 1982, Mud transportaion on a steep shelf, Rio de la Plata shelf, Puerto Rico: Geo. Marine Letters, v. 2, p. 71-75.

Hayes, M. O., 1967, Hurricanes as geologic agents, south Texas coast: Am. Assoc. Petroleum Geologists Bull., v. 51, p. 937-956.

Howard, J. D., and Reineck, H. E., 1981, Depositional facies of high-energy beach-to-offshore sequence, comparisons with low-energy sequence: Am. Assoc. Petroleum Geologists Bull., v. 65, p. 807-830.

Kreisa, R. D., 1981, Storm-generated sedimentary structures in subtidal marine facies with examples from the middle and upper Ordivician of southwestern Virginia: Jour. Sed. Petrology, v. 51, No. 3, p. 823-848.

Kreisa, R. D., and Bambach, R. K., 1982, The role of storm processes in generating shell beds in Paleozoic shelf environments, in Einsele, G., and Seilacher, A., eds., Cyclic and event stratification: Springer-Verlag, New York, p. 200-220.

Kumar, N., and Sanders, J. E., 1976, Characteristics of shoreface storm deposits: modern and ancient examples: Jour. Sed. Petrology, v. 46, p. 145-162.

Lavelle, J. W., Swift, D. J. P., Gadd, P. E., Stubblefield, W. L., Case, F. N., Brashear, H. R., and Haff, K. W., 1978, Fair weather and storm sand transport on the Long Island, New York, inner-shelf: Sedimentology, v. 25, p. 823-842.

Lighty, R. G., Macintyre, I. G., and Stuckenrath, 1982, Acropora palmata reef framework, a reliable indicator of sea level in the western Atlantic for the past 10,000 years: Coral Reefs 1:125-30.

Lopez, M. A., 1964, Floods at Toa Alta, Toa Baja, and Baja, and Dorado, Puerto Rico: U.S. Geol. Survey Hydrologic Investigations Atlas H.A.-128.

Lopez, M. A., and Colon-Dieppa, E., 1973, Magnitude and frequency of floods in Puerto Rico: Co-operative Water Resource Investigation Data Release PR-9, 63 p.

MacDonald, K. B., 1976, Paleocommunities: toward some confidence limits, in Scott, R. W., and West, R. R., eds., Structure and classification of paleocommunities: Dowden, Hutchinson, and Ross, Inc., Stroudsburg, Pa., p. 87-106.

Morelock, J., Schwartz, M. L., Hernandez-Avila, M., and Hatfield, D. M., 1985, Net shore-drift on the north coast of Puerto Rico: Shore and Beach, v. 53, No. 4, p. 16-21.

Pilkey, O. H., Trumbull, J. V. A., and Bush, D. M., 1978, Equilibrium shelf sedimentation, Rio de la Plata Shelf, Puerto Rico: Jour. Sed. Petrology, v. 48, No. 2, p. 389-400.

Pilkey, O. H., Fierman, E. I., and Trumbull, J. V. A., 1979, Relationships between physical condition of the carbonate fraction and sediment environemnts, northern Puerto Rico shelf: Sediment. Geol., v. 24, p. 283-90.

Pilkey, O. H., Bush, D. M., and Rodriguez, R. W., 1984, Storm sedimentation; North shelf of Puerto Rico, in Park, Y. A., Pilkey, O. H., and Kim, S. W., eds., Proceedings of Korea-U.S. Seminar and Workshop on Marine Geology and Physical Processes of the Yellow Sea: Korea Institute of Energy and Resources, Seoul, South Korea, p. 242-259.

Reineck, H. E., and Singh, I. B., 1972, Genesis of laminated sand and graded rhythmites in storm sand layers of shelf mud: Sedimentology, v. 18, p. 123-128.

Schneidermann, N., Pilkey, O. H., and Saunders, C., 1976, Sedimentation on the Puerto Rico insular shelf: Jour. Sed. Petrology, v. 46, p. 167-173.

Secretary of the Army, 1962, San Juan, Puerto Rico, beach erosion control study, 87th Congress, 2d session, House Document No. 575: Washington, D.C., U.S. Government Printing Office.

Seilacher, A., 1982, General remarks about event deposits, in Einsele, G., and Seilacher, A., eds., Cyclic and event stratification: Springer-Verlag, New York, p. 161-173.

Shinn, E. A., Hudson, J. H., Halley, R. B., Lidz, B., Robbin, D. M., and Macintyre, I. G., 1981, Geology and sediment accumulation rates at Carrie Bow Cay, Belize: Smithsonian Contributions to the marine sciences, No. 12, p. 63-75.

Swift, D. J. P., 1976, Coastal sedimentation, in Stanley, D. J., and Swift, D. J. P., eds., Marine sediment transport and environmental management: New York, John Wiley and Sons, p. 235-310.

Swift, D. J. P., Stanley, D. J., and Curray, R. J., 1971, Relict sediments on continental shelves: a reconsideration: Jour. Geology, v. 79, p. 323-346.

Thayer, C. W., 1979, Biological bulldozers and the evolution of marine benthic communities: Science, v. 203, p. 458-460.

Warme, J. E., Elkdale, A. A., Elkdale, S. G., and Peterson, C. H., 1976, Raw material of the fossil record, in Scott, R. W., and West, R. R., eds., Structure and classification of paleo-communities: Dowden, Hutchinson, and Ross, Inc., Stroudsburg, Pa., p. 143-169.

Western Geophysical of America, 1974, Offshore geophysical investigations for siting of a nuclear power station on Puerto Rico, Appendix 2.5A, Geophysical reconnaissance and exploration of Puerto Rico: Final report for the Puerto Rico Water Resources Authority.

Western Geophysical Corporation, 1975, Shallow water bathymetric, sonar and seismic investigations for siting the north coast nuclear plant 1 on Puerto Rico: Appendix 2.5R, Amendment 27.

Wood, E. D., Youngbluth, M. J., Nutt, M. E., Yoshioka, P., and Canoy, M. J., 1975a, Tortuquero Bay environmental studies: Puerto Rico nuclear center, Publication #181, 227 p.

Wood, E. D., Youngbluth, J. J., Nutt, M. E., Yeaman, M. N., Yoshioku, P., and Canoy, M. J., 1975b, Punta Manati environmental studies: Puerto Rico Nuclear Center, Publication #182, 225 p.

FACIES ANALYSIS AND RESERVOIR ZONATION OF A

CRETACEOUS SHELF SAND RIDGE:

HARTZOG DRAW FIELD, WYOMING

Robert S. Tye,[1] Vishnu Ranganathan,[2] and W. J. Ebanks, Jr.[3]

[1]Coastal Studies Institute, Louisiana State University, Baton Rouge, Louisiana 70803

[2]Department of Geology, Louisiana State University, Baton Rouge, Louisiana 70803

[3]ARCO Exploration and Production Research, 2300 West Plano Parkway, Plano, Texas 75075

Abstract

Geologic analysis of the Hartzog Draw Field pilot area indicates that reservoir behavior is predominantly controlled by the processes of deposition and the diagenetic history of the genetic facies composing the reservoir. Central-bar, bar-margin, and interbar facies of the Cretaceous Shannon Sandstone were deposited as an elongate shelf sand ridge in the Cretaceous seaway of northern Wyoming.

A typical vertical sequence in the Hartzog Draw Field coarsens upward and consists of rippled to bioturbated, very fine-grained interbar sandstone interlayered with trough cross-bedded and rippled medium- to fine-grained bar-margin sandstone and overlain by cross-bedded and rippled medium- to fine-grained central-bar sandstone. The bar-margin facies generally has a sharp basal contact and exhibits an upward trend from trough cross-beds to small-scale cross-bedding and ripples that grade into the central bar. Interfingering contacts of central bar with shalier bar-margin and interbar facies disrupt the vertical and horizontal reservoir continuity. The bar-margin facies is best developed on the steeply sloping eastern margin of the field and interfingers westward into thicker central-bar sandstone. Westward of the northwest-southeast axis, the interbar facies and shelf shales split the reservoir into three horizons, thus creating thinner and less laterally extensive sand lenses at the westward extent of the field.

Core samples were analyzed by thin-section petrography, SEM (scanning electron microscope), and XRD (X-ray diffraction) to assess primary and secondary controls on porosity. Shannon sands are feldspathic-litharenites (60% quartz, 15% feldspar, and 25% lithic rock fragments). Primary controls on porosity are sediment texture and the amount of labile grains (glauconite, chert, and rock fragments). Secondary properties that modify porosity include carbonate cementation, sedimentary structures, dissolution of carbonate cement, and occlusion of secondary porosity by compaction and formation of quartz and clay cements. Core plug permeabilities and porosities observed in thin sections for cross-bedded central-bar and bar-margin sandstone range from 1 to 40 md and 2.5% to 12%, respectively. Thinner-bedded and shalier bar-margin and interbar facies have permeabilities ranging from less than 1 md to 6 md and thin-section porosities from 0% to 2%. Lower porosity and permeability values are caused by a large increase in detrital clay content of these units.

Flow units represent portions of the reservoir that have similar fluid-flow properties. Integration of stratigraphic, sedimentologic, and petrophysical data into reservoir flow units improves predictions of trends in reservoir quality. Within the flow units, geologic features that greatly influence fluid flow are (1) stratification, (2) shale laminations, (3) cement, and (4) cross-bedding.

Introduction

Hartzog Draw Field is located approximately 48 km southwest of Gillette, Wyoming, in Campbell and Johnson counties (Fig. 1). Production is from the Upper Cretaceous Shannon Sandstone member of the Cody (Pierre) Shale. The reservoir sandstone occurs at depths between 2804 and 2896 m, at an average depth of 2865 m, and the field trends northwest-southeast. It is presently 43.2 km long and ranges from 1.6 to 6.4 km wide (Fig. 2). As of October 1983, 115 wells produced 8000 barrels/day of oil and 4500 MCF/day of gas from a gross thickness of 6 m (Hunt and Hearn, 1981). Descriptions of field development and production history are presented by Hunt and Hearn (1981).

Deposition of the Shannon Sandstone as multiple sand ridges on the submarine shelf near the western margin of the Cretaceous Interior seaway occurred during late Cretaceous (Campanian) time (Gill and Cobban, 1969). Paleogeographic reconstruction of the Powder River basin and paleontologic data from shales above and below the Shannon indicate that deposition occurred approximately 160 km from the shoreline in water as deep as 18-91 m (Gill and Cobban, 1969; Martinsen and Tillman, 1979). The Shannon Sandstone is enclosed by the Cody Shale,

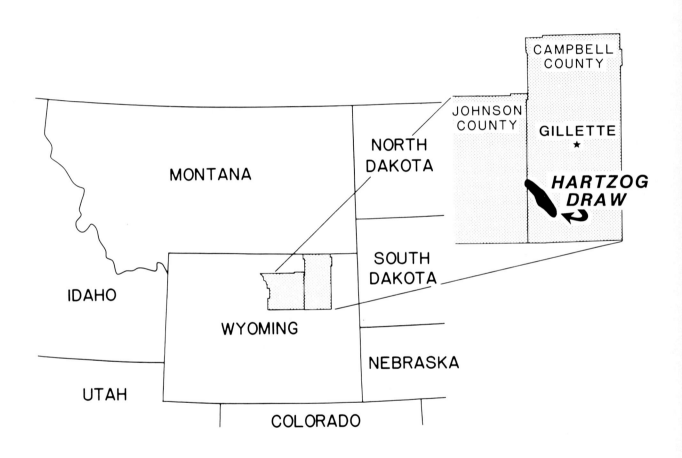

Figure 1. Location of Hartzog Draw Field, Campbell and Johnson
counties, Wyoming.

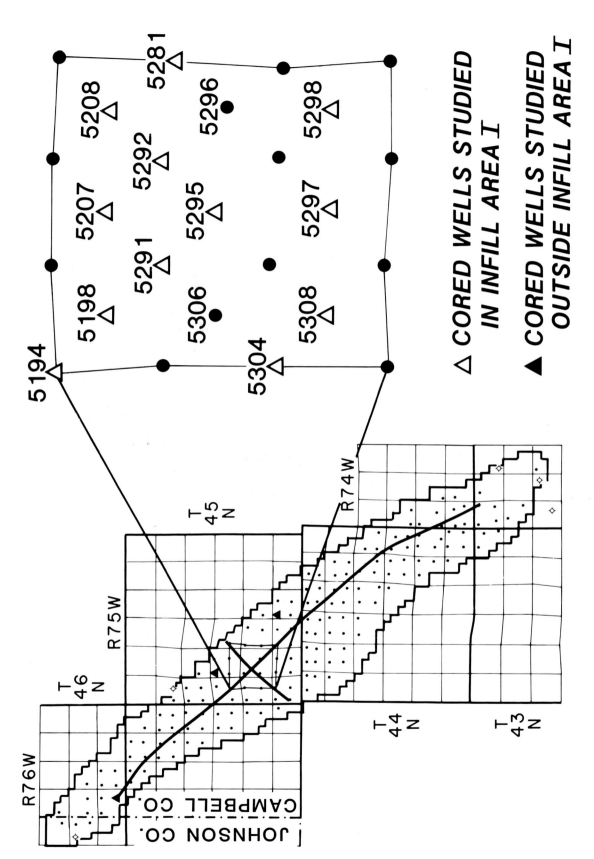

Figure 2. Hartzog Draw Field and the infill area on which this study is focused. Numbers are well numbers referred to in the text. Note locations for cross sections in Figures 4 and 18.

△ CORED WELLS STUDIED IN INFILL AREA I

▲ CORED WELLS STUDIED OUTSIDE INFILL AREA I

which has a regional dip of 2° southwest. This combination of a very slight structural dip with no faults and a shale envelope around the sandstone has resulted in an excellent stratigraphic trap for hydrocarbons. Geologic interpretations of the Shannon Sandstone in Hartzog Draw have been made by Tillman and Martinsen (1979). Oil and gas accumulations in similar depositional and stratigraphic settings have been described in the Viking Formation (Evans, 1970), the Cardium Formation (Berven, 1966), the Hygiene Member of the Pierre Shale (Porter, 1976), and the Sussex member of the Pierre Shale (Berg, 1975; Hobson et al., 1982).

The effects of areal variation of reservoir properties became evident during the primary production period in Hartzog Draw. A contour map of cumulative oil production through 1980 per well-bore porosity-foot (Fig. 3) shows a marked disparity between greater production in the eastern portion of the field and lower production in the western. This production trend is related to the reservoir geology and will be discussed further.

Water flooding began in 1981 and a 9-mi^2 pilot area was chosen for a detailed reservoir study in anticipation of a later carbon dixoide flood (Fig. 2). Nineteen cores (15 within the pilot area and 4 on the north and east margins) with an average length of 18 m were analyzed for sedimentologic and diagenetic characteristics. Four cores (5341, 5308, 5298, 5208) were oriented with respect to magnetic north to establish the predominant directions of dip of cross-bed sets within the sandstone and to evaluate this possible influence on directional permeability.

The comprehensive geological evaluation of the Hartzog Draw pilot area presented here is intended to expand on the earlier work by

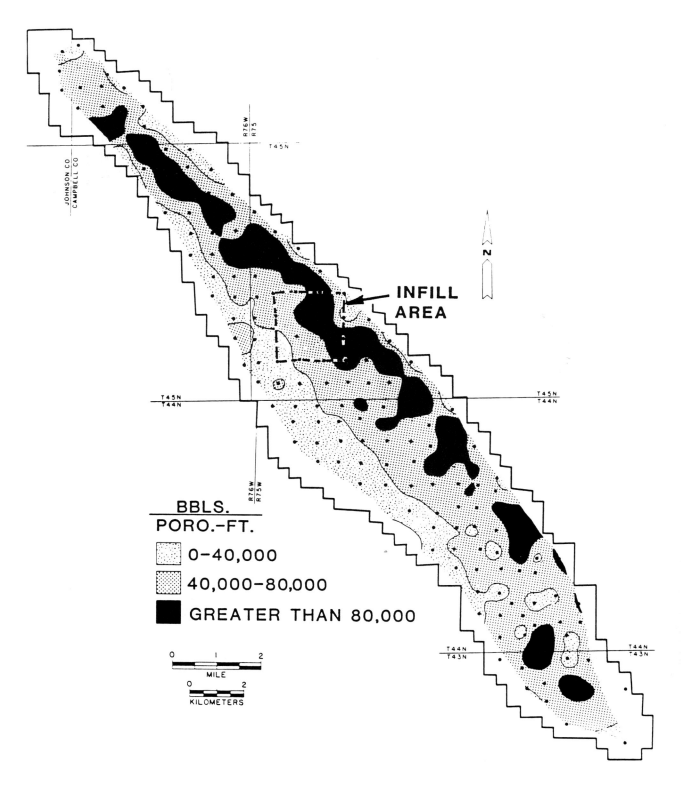

Figure 3. Map of cumulative production of Hartzog Draw Field expressed as barrels per porosity foot through 1980. "Better quality" sand as expressed by this map does coincide with the trend of thicker sandstone as shown by the gross sandstone thickness map (Fig. 15).

174

Martinsen and Tillman (1979) and to provide a detailed geologic description of the area for improved reservoir simulation. Emphasis has been placed on the stratigraphic continuity of reservoir sandstone, variability of reservoir and nonreservoir rocks, the depositional and diagenetic controls on porosity and permeability, and recognition of potentially troublesome minerals that may cause formation damage or affect log analysis.

Geologic Facies and Reservoir Sandstone Geometry

The Shannon Sandstone of Hartzog Draw Field contains four sandstone and two siltstone-to-shale lithofacies arranged in the form of three stacked sandstone lenses. The lenses have been designated as the upper (U), middle (M), and lower (L) reservoir lenses (Fig. 4), and production is encountered in each lens. Silty shale beds separate these lenses in most areas of the field, but in some areas they appear to be in contact. Tillman and Martinsen (1979, 1984) first recognized this tripartite character of the Hartzog Draw Field and outlined the major depositional environments of these three lenses. The "quality" of the reservoir rocks in each lens is related both to the depositional environment of the sandstone and to modifications of the original pore systems by diagenesis. Field production is not a simple relationship between gross sandstone thickness and oil recovery.

In this paper, descriptions of geologic facies are patterned after those of Porter (1976), Martinsen and Tillman (1979), and Boyles and Scott (1982), whose studies also pertained to Cretaceous shelf sandstones. The primary sedimentary facies are distinguished on the basis of sedimentary structures and lithologic characteristics. Figure 5, a composite log from two wells in the field, illustrates the common

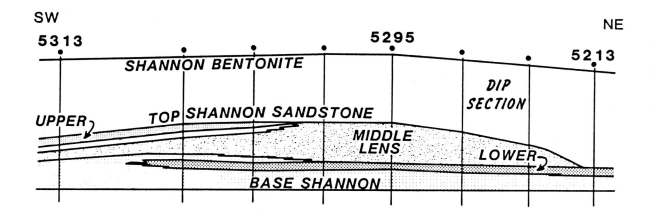

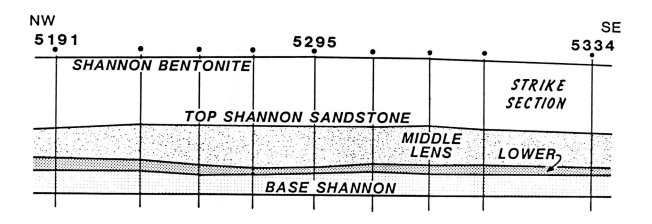

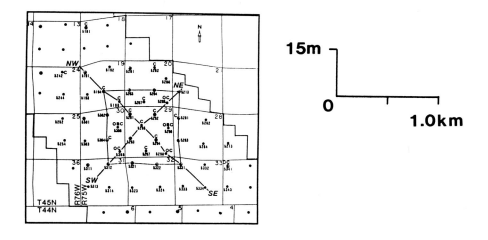

Figure 4. Stratigraphic cross sections of the upper, middle, and lower Shannon Sandstone lenses. Note asymmetry of both the Shannon thickness and the distribution of the lenses in the dip (SW-NE) cross sections. See Figure 2 for locations.

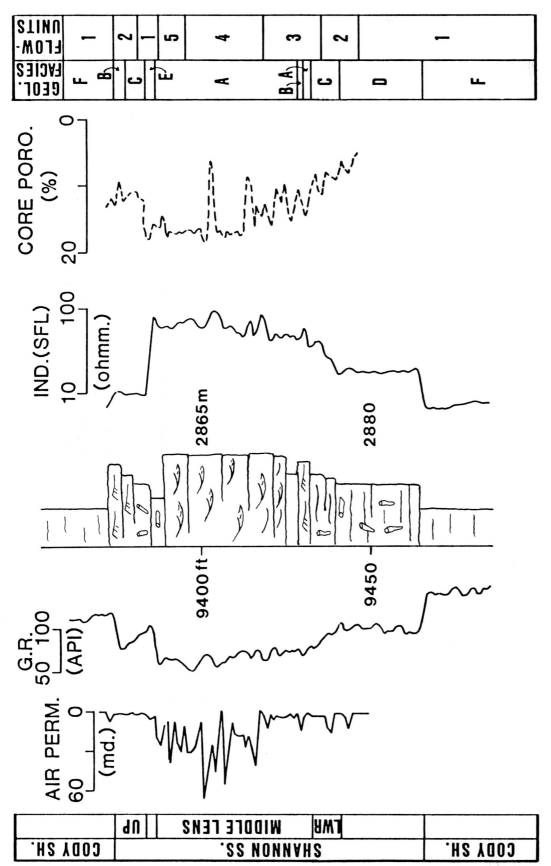

Figure 5. Composite reservoir quality profile, based on cores and logs of wells 5295 and 5308, showing relationships of lithology to log character and porosity and permeability. Also shown are subdivisions of the Shannon Sandstone (sand lenses, facies, and flow units). Note that flow units and facies coincide only in part, owing to emphasis given petrophysical data in definition of flow units.

vertical arrangement of the six lithologic facies, their log character, the typical porosities and permeabilities of each facies, and facies relationships to the Shannon sand lenses.

The coarsest-grained lithofacies within the Shannon Sandstone have been designated facies A-C and are briefly described below in the context of their occurrence in the three sandstone lenses. Facies A, the central-bar facies, which comprises most of the middle lens (Figs. 4, 6), is cross-bedded, medium- to fine-grained sandstone having scattered clasts (gravel-size fragments) of sideritic mudstone and only minor laminae of shale. The central-bar facies becomes slightly more fine grained and shaly downward. In the lower one-half to one-fourth of the middle lens, the central-bar facies is interbedded with the bar-margin facies (facies B and C).

Facies B, the bar-margin facies, type 1, consists of cross-bedded and rippled coarse- to medium-grained sandstone that has clasts of shale and sideritic mudstone. This facies occurs at the top of the upper and middle lenses and in thin interbeds below the central-bar facies (Figs. 4, 6). Facies B is more common on the eastern side of the field than on the western.

Facies C, the bar-margin facies, type 2, is composed of laminated and ripple-bedded, shaly sandstone, in some instances being moderately burrowed. This facies occurs in the lower parts of the upper and middle lenses and comprises most of the lower lens (Figs. 4, 6).

Fine-grained, relatively poorly sorted sandstone, siltstone, and shale form breaks between the sandstone lenses and enclose the reservoir sandstone. Beneath the Shannon Sandstone is a transitional unit of facies D, the interbar facies. Facies D sandstone is fine-grained

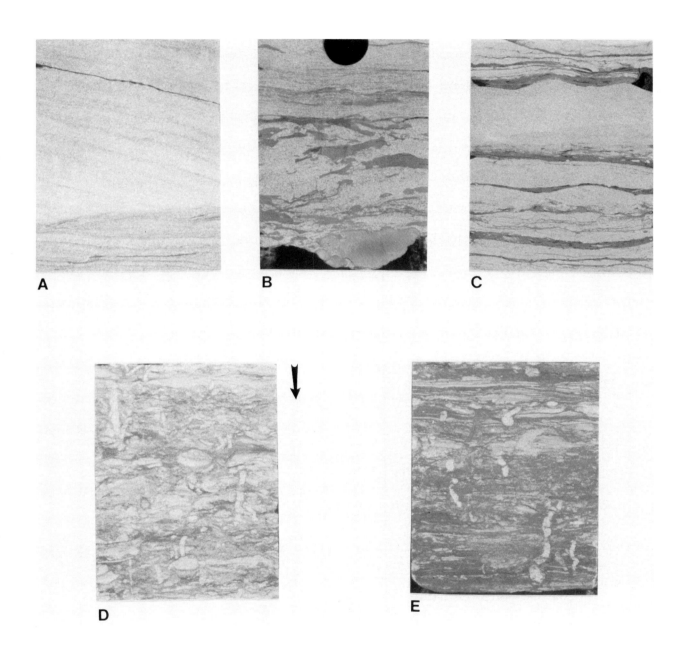

Figure 6. Representative slabbed core samples of the five lithofacies in Hartzog Draw Field. Each core sample is 3 in. wide, and the arrow indicates the downhole direction.

and ripple-bedded, and contains numerous shale laminations and burrows. The upper lens is separated from the middle lens by a thin, intensely burrowed, shaly siltstone of Facies E, the bioturbated siltstone facies (Figs. 4, 6). Beneath the lower lens and above the upper lens is Facies F, the shelf shale facies. The shelf shale facies is the Cody Shale that encases the Shannon Sandstone complex.

In descriptions of the facies, sedimentary parameters are expressed as proportions (percentage by volume), according to the visually estimated relative abundance of each feature in cores. For example, the total quantities of sandstone, siltstone, and shale must equal 100%. Likewise, the total of physical and biogenic sedimentary structures must equal 100%. A summary of sedimentary structures, lithology, and reservoir properties of each facies is given in Table 1.

Lithofacies

Facies A. Facies A, a cross-bedded, medium-grained sandstone, has the greatest measured porosity and permeability of all of the facies. It ranges in thickness from 3.1 to 14.0 m and consists of 82%-89% coarsening-upward, fine- to medium-grained, well-sorted sandstone having an average grain size of 0.3 mm. Core plug porosities range from 2.9% to 22.5% and average 14.8%. Permeabilities range from 0.01 to 143.00 md and average 16.3 md. Megascopic controls on porosity include localized concentrations of shale clasts and laminations. Early calcite cement and siderite cement occlude porosity in zones 2-30 cm thick. Shale content in the sandstone ranges from less than 1% to 10%, and laminated shale, occurring as drapes over rippled beds, is the dominant form. Shale rip-up clasts are also common. Glauconite is common on bedding planes. Siderite can occur as thin sideritic mud

Table 1. Lithology, sedimentary structures, and reservoir characteristics of facies A–F. Reservoir characteristics are average properties measured in cores.

	Facies A	Facies B	Facies C	Facies D	Facies E	Facies F
Lithology	Fine to medium-grained sandstone (0.2-0.3 mm); sparse shale and siderite laminations and clasts.	Fine to medium-grained sandstone (0.2-0.4 mm); common shale and siderite clasts.	Fine-grained sandstone (0.2 mm); thin shale laminations; minor siderite.	Very fine-grained sandstone (0.1-0.2 mm); moderate shale laminations.	Siltstone with fine-grained sandstone (0.15 mm); common shale laminations.	Silty shale with trace of fine-grained sandstone (0.15 mm).
Physical Sedimentary Structures	Trough and planar cross-bedding; current ripples; horizontal laminations; soft-sediment deformation.	Trough cross-bedding; current ripples; soft-sediment deformation.	Current ripples; trough cross-bedding; wavy bedding; horizontal laminations.	Wavy and lenticular bedding; current ripples; horizontal laminations.	Rare lenticular and wavy bedding.	Horizontal laminations; starved ripples.
Biogenic Sedimentary Structures	Rare burrowing.	Rare burrowing.	Common burrowing; moderate diversity; rare bioturbation.	Abundant burrowing; high diversity; rare bioturbation.	Almost total bioturbation.	Moderate burrowing; moderate diversity; rare bioturbation.
Reservoir Characteristics	Porosity = 14.8% Permeability = 16.3 md Water saturation = 15%	Porosity = 12.9% Permeability = 10.7 md Water saturation = 18%	Porosity = 11.3% Permeability = 3.5 md Water saturation = 25%	Porosity = 8.6% Permeability = 1.4 md Water saturation = 35%	Porosity = 9.8% Permeability = 1.9 md Water saturation = 50%	

laminations, rip-up clasts, nodular concretions, and cement. Zones containing siderite clasts in excess of 10% have often undergone early siderite cementation. Tightly cemented zones may be as thick as 15 cm, but their lateral extent is not known. If they are similar to comparable features in outcrops of Shannon Sandstone, the lenses of sideritic mudstones may be only 30-40 m in areal extent.

Trough cross-bedding is the most conspicuous type of sedimentary structure in facies A. Planar cross-bedding is apparent in less than one-third of the cores described, and when present, it is concentrated at the base and near the top of facies A. Thickness of bed sets typically ranges from 3 to 45 cm. Current ripples and current-modified wave ripples are very common in facies A. Extensively rippled zones are usually concentrated at the base of the sandstones, whereas trough cross-beds are more common in the upper parts. Bioturbation in excess of 75% burrows is rare. Burrows are concentrated in zones of interlaminated rippled sandstone and shale. Paleotransport directions measured from trough and planar cross-beds in oriented cores indicate sediment transport to the southeast (131° to 169°) at an acute angle to the long axis of the reservoir. A large proportion of trough and planar cross-beds and asymmetrical ripples indicates that sediment deposition was accomplished by unidirectional currents that were occasionally influenced and enhanced by storm-generated currents.

Gamma-ray log signatures of facies A reflect coarsening and improved sorting in the upper portion of the sandstone (Fig. 5). Core-to-log correlations indicate that numerous thin shale laminations and thin beds of shale clasts, siderite, glauconite, and authigenic clay influence the gamma-ray curve. Thus, the shalelike inflections on the gamma-ray log probably do not represent laterally continuous shale layers. Sharp, low-porosity inflections on density-porosity logs correlate reasonably well with beds of siderite or calcite cement, although the actual thickness of the cemented interval is often exaggerated by the log.

Facies B. Facies B, cross-bedded medium- to coarse-grained sandstone, ranges in thickness from 15 cm to 3.75 m and consists of 75%-96% medium- to coarse-grained sandstone (average grain size, 0.3 mm). Core porosity in facies B ranges from 2% to 18.8% and averages 12.9%. Permeability ranges from 0.1 to 73.0 md and averages 10.7 md. Shale clasts and laminations make up 5%-25% of facies B. Rip-up clasts are the dominant form of shale. Siderite is more abundant (5%-30% by volume) in facies B than in facies A. Rip-up clasts and replacement nodules of siderite create intraformational conglomerates (Fig. 6). Glauconite concentrated on bedding surfaces can exceed 30% of the grain volume of these sandstones.

Trough cross-bedding comprises 10%-50% of the physical structures. Planar cross-beds are more abundant than in facies A, and bed sets average only 10 cm thick. Current ripples and ripples superimposed on troughs account for 10%-70% of the physical sedimentary structures. Several factors, including coarser texture, common rip-up clasts, dominance of trough cross-bedding, and low biogenic activity, indicate that facies B was deposited under the highest energy conditions. That high-energy conditions occurred episodically is indicated by poorer sorting (relative to facies A), the thinness and limited lateral extent of the sand layers, the presence of small-scale current ripples, and rare burrowed shale laminae.

Increased quantities of shale, siderite, and glauconite often cause facies B sandstones to appear as shaly intervals on gamma-ray and porosity logs. However, core analyses indicate that most of the individual sandstone layers are good-quality reservoir rock, except where they are cemented by carbonate.

Facies C. Sandstones of facies C, laminated and rippled, have more current and wave-ripple bedding, shale laminations, and burrowing than sandstones of facies A or B (Fig. 6). The thickness of these fine-grained sandstones (average grain size, 0.2 mm) ranges from 24 cm to 3-9 m. Shale laminations, and rarely shale clasts, compose 10%-30% of facies C. Siderite is rare but some thin 1.5-cm burrowed beds are present. Facies C has low core porosity and permeability because of a high proportion of burrowing and the occurrence of shale laminations. Core porosity ranges from 4.0% to 17.4% (average, 11.3%), and permeability ranges from 0.01 to 16.1 md (average, 3.5 md).

Interfingering of facies A, B, and C produces an overall "cleaning"-upward character in the sandstone reflected in the gamma-ray log (Fig. 5). The gamma-ray log also illustrates the gradation of facies C into the underlying, very shaly sandstone and shale (Fig. 5).

Facies D. Fine-grained sandstone (average grain size, 0.15 mm) makes up from 40% to 80% of facies D, a burrowed shaly sandstone. Siltstone content is about 5%, and shale content varies from 10% to 40%. Rippled sandstone occurs in equal proportion to wavy and lenticular-bedded sandstone. Bioturbation averages 25%. The thinly interbedded and churned nature of sandstone and shale in facies D produces a slightly irregular gamma-ray curve intermediate to the curve for facies F (shale) and the cleaner, better-sorted facies A and B sandstones. Like facies C, facies D has low core porosity and permeability. Core porosity ranges from 3.2% to 15.2% (average, 8.6%), and permeability ranges from 0.01 to 34.0 md (average, 1.4 md).

Facies E. Facies E, bioturbated siltstone, was observed in only two cores from the westward extremity of the study area. Fine-grained sandstone (average grain size, 0.15 mm) makes up 20%–26% of this facies; siltstone constitutes 44%–50% and shale 30%. Siderite is present from trace amounts to 11% and occurs as clasts, replacement nodules, laminations, and cement. At least 75% of facies E is bioturbated, but thin, discontinuous, wavy and lenticular beds of silty sandstone and small-scale rippled sandstone occur. Facies E has low core porosities and permeabilities because of homogenization by burrowing. Core porosities range from 5.8% to 12.9% (average, 9.8%), and permeabilities vary from 0.04 to 5.5 md (average, 1.9 md).

Facies F. Overlying and underlying shales envelope the Shannon Sandstone, creating an impermeable seal around the reservoir. Facies F consists of 84% shale, 13% siltstone, and only 3% fine-grained sandstone. It is evident from the sharp upper sand/shale contact of the Shannon Sandstone that deposition or reworking of large quantities of sand ceased fairly abruptly, thus ending the period of shelf sand-ridge growth.

Vertical Sequences

The vertical distribution of lithofacies follows a predictable pattern determined by the core's geographic position relative to the northwest-southeast axis of the reservoir. An east-west transect of four cores in the pilot area (H.D.U. 5202, 5281, 5295, and 5308) illustrates the occurrence of facies (Fig. 7). Sandstone thickness, quality, and vertical arrangement of facies vary greatly in these four cores.

Core 5202. The easternmost core, 5202, exhibits a well-developed section with 6.35 m of facies D (burrowed, shaly sandstone), resting conformably on facies F (Fig. 7). Abruptly overlying facies D is 1.7 m of interbedded facies C (laminated and rippled sandstone) and Facies B (cross-bedded coarse- and medium-grained sandstone). Facies B sandstone accounts for most of the production in this thin interval (Fig. 8). Sandstone of facies B along the eastern edge of the field includes a large proportion of shale clasts, siderite, and glauconite.

Core 5281. Well 5281 is located on the eastern margin of the infill area, somewhat more than 1 km to the southeast and structurally down-dip of well 5202. A thin 60-cm interval of facies C sandstone represents a gradational change between underlying facies D and a thick (4.3-m), well-developed facies B interval (Fig. 7). In core 5281, 4.31 m of facies A occurs between two intervals of facies B (Fig. 9). Facies F caps the sequence of sandstones.

Core 5295. Core 5295, located on the axis (center line) of the Hartzog Draw Field, contains the maximum thickness of sandstone (Figs. 7, 10). Basal facies D deposits coarsen upward into facies C, which is

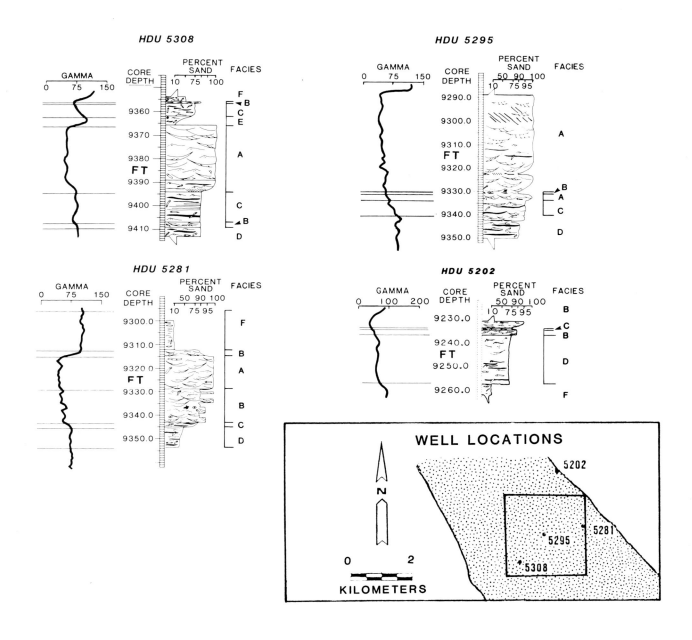

Figure 7. Vertical sequences from four cores across the reservoir illustrating the facies variability in the Shannon Sandstone at Hartzog Draw.

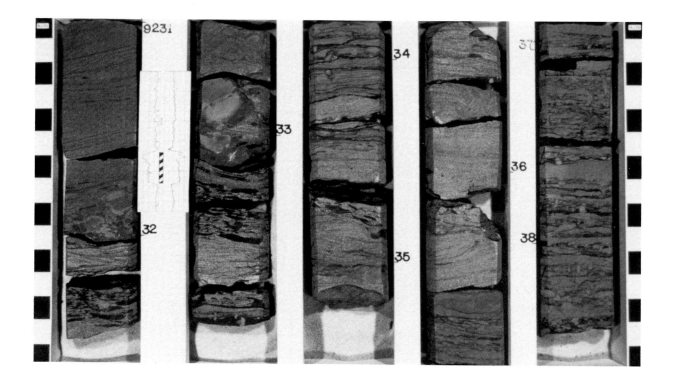

Figure 8. Representative slabbed samples of core 5202. Compare with the vertical sequence in Figure 7 to note characteristics of facies B, C, D, and F. Depths are in feet. Facies B and C (9231 to 9237 ft); facies D (9237 to 9257 ft); facies F (9257 to 9262 ft).

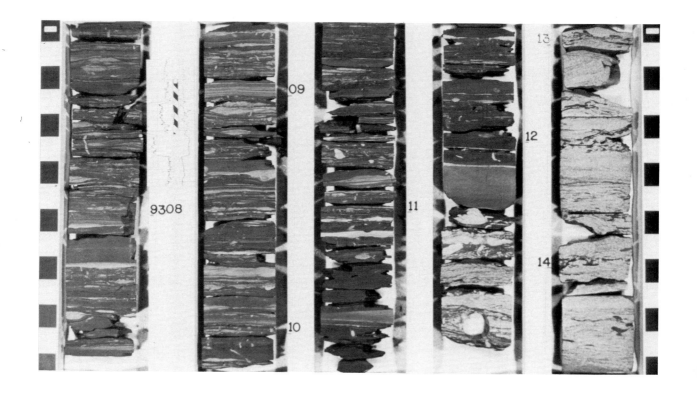

Figure 9. Slabbed samples of core 5281. Approximately 4 m of facies A sandstone (9315 to 9328 ft) is overlain and underlain by facies B. Note the sharp basal contact of the overlying facies F shale. Depths are in feet.

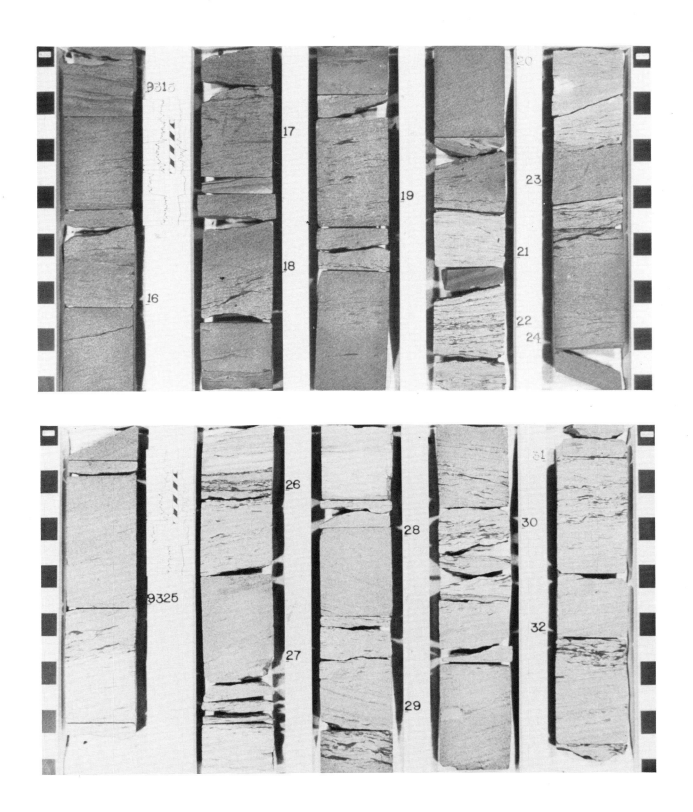

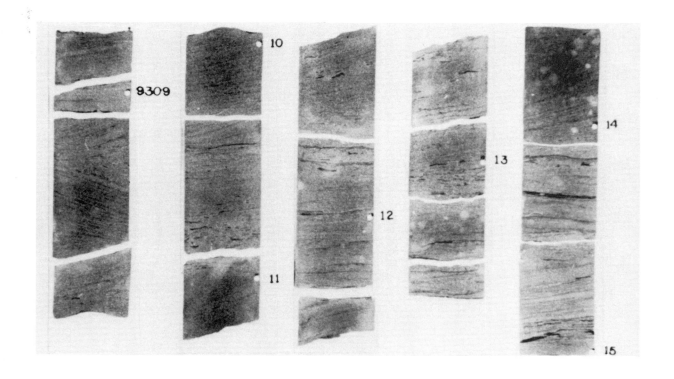

Figure 10.   Slabbed samples of core 5295 illustrating the well-sorted and cross-bedded nature of the facies A sandstone (9309 to 9321 ft).   Compare facies A with the underlying burrowed shaly facies D sandstone (9342 to 9347 ft). Dark coloration in facies A is oil stain.   Depths are in feet.

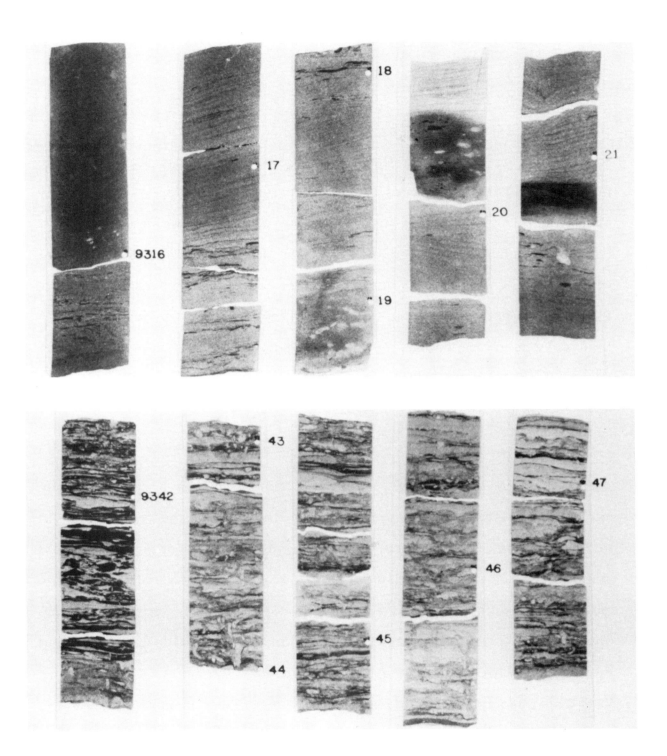

successively overlain by thin intervals of facies A and facies B. Facies A (15.7 m) occurs at the top.

Core 5308. Core 5308, located 1.17 km west of the northwest-southeast reservoir axis, exhibits the most variability in vertical arrangement of facies among the cores studied (Fig. 11). Well-developed facies B and C sandstones overlie facies D deposits. Gradationally overlying facies B and C is 8.5 m of coarsening-upward facies A. Facies A is abruptly capped by 2.77 m of fine-grained facies E siltsone, which becomes more sandy upward. This rapid change from facies A to facies E reflects a sudden decrease in energy during deposition. A final episode of deposition by higher-energy currents is evidenced by 15 cm of facies B that forms a thin, coarse-grained sandstone cap (the upper lens) on the facies E siltstone.

Stratigraphy

Within the study area, the lower Shannon lens is a relatively thin wedge of sandstone that ranges from 4 m thick along the eastern margin to zero along the western (Figs. 4, 12). A northeast-southwest trending zone of thin sandstone (1 m) divides the lower lens into two lobes with thicknesses as great as 4-5 m. The lower lens extends an undetermined distance east of the present field boundary and interfingers with shale to the west.

The Shannon Sandstone in Hartzog Draw Field is dominated by the middle lens (Figs. 4, 13). From the trend of thickest sandstone (14-15 m) near the northwest-southeast trending axis of the field, the middle lens thins abruptly eastward and more gradually westward to beyond the present field boundary. Minimum thickness of this lens on the

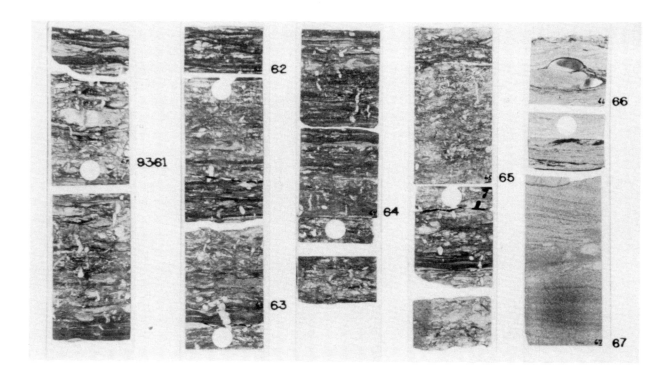

Figure 11. Slabbed samples of core 5308 illustrating the stacking and repetition of facies B (9356.0 to 9356.5 ft), C (9356.5 to 9362.0 ft), E (9362.0 to 9365.5 ft), A (9365.5 to 9394.0 ft), C (9394.0 to 9407.0 ft), B (9407.0 to 9409.0 ft), and D (9409.0 to 9413.0 ft). Facies E siltstone separates the upper and middle sandstone lenses.

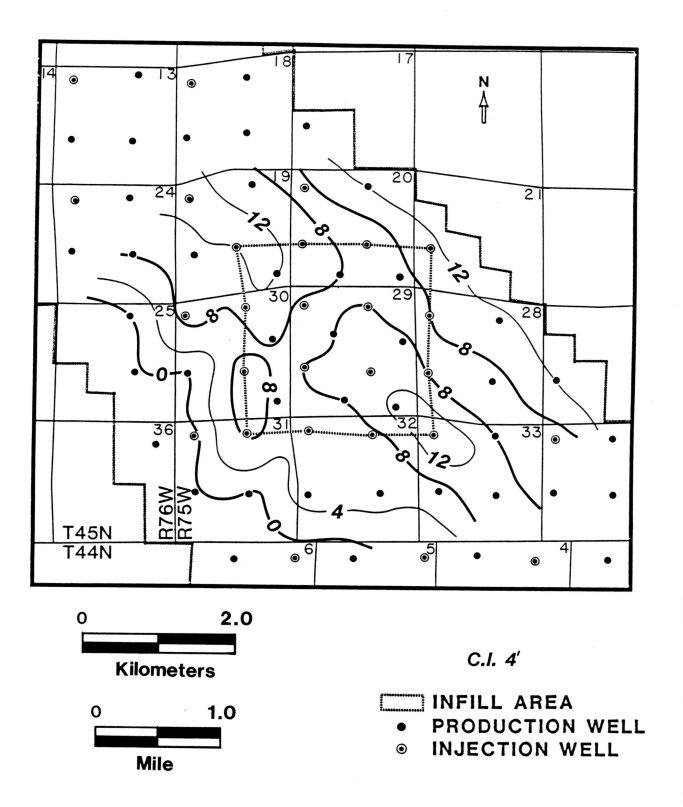

Figure 12. Thickness of the lower lens of the Shannon Sandstone within and around the study area. Contour interval is in feet.

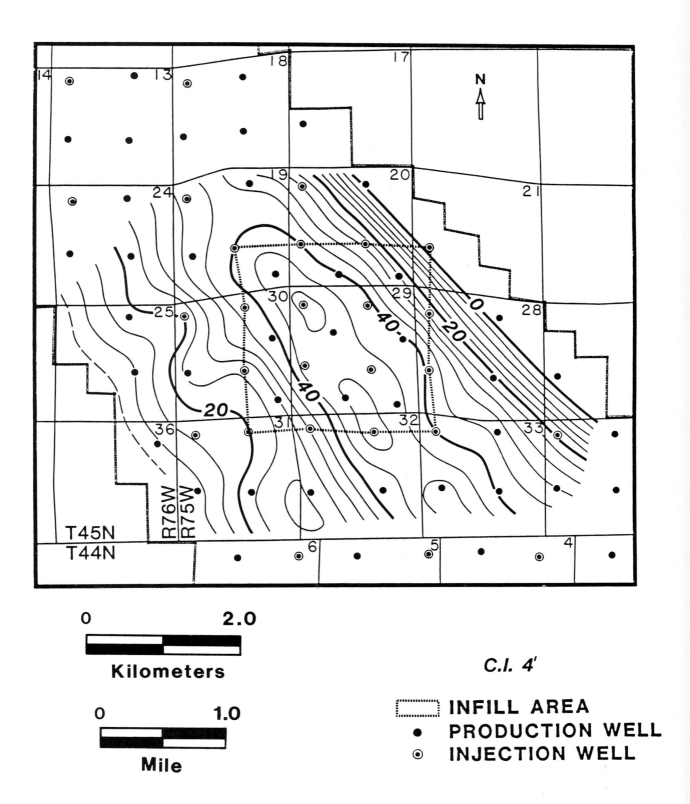

Figure 13. Thickness of the middle lens of the Shannon Sandstone within and around the study area. Contour interval is in feet.

western edge of the field is 2.4 m. In plan view, two areas of thicker sand near the axis correspond approximately to a similar thickening trend in the underlying lower lens. The trend of the eastern margin is fairly straight, whereas the western margin is more irregular or scalloped in plan view.

The upper lens of the Shannon is a fairly simple wedge within the infill area (Figs. 4, 14). From a pinch-out near the axis of the field and near the thickest trend in the underlying middle lens, this upper lens thickens gradually to 2.5 m and then thins to 1.2 m toward the southwest. This upper lens lacks the irregularities or lobes of thicker and thinner sandstone on the scale of the lower two lenses. Thicknesses of these three lenses are shown in three isopach maps (Figs. 12-14), and their aggregate gross thickness is shown in Figure 15.

The upper and lower reservoir lenses consist almost totally of bar-margin (facies C) and interbar sandstones (facies D), whereas the better-sorted and coarser-grained bar margin (facies B) and central bar compose the middle lens. In a strike-oriented cross section across the entire field, the greatest thickness of sandstone occurs in the north-central portion of the field. Central-bar sandstone up to 10 m thick and interfingering bar-margin deposits characterize the middle lens, which overlies laterally continuous bar-margin sandstone of the lower lens (Fig. 16). Facies associations are slightly more complex along depositional dip (Fig. 17). Along the basinward margin of the shelf ridge, bar-margin sandstone forms the lower lens and interfingers with the central bar to form a coarse-grained, poorly sorted nose on the ridge.

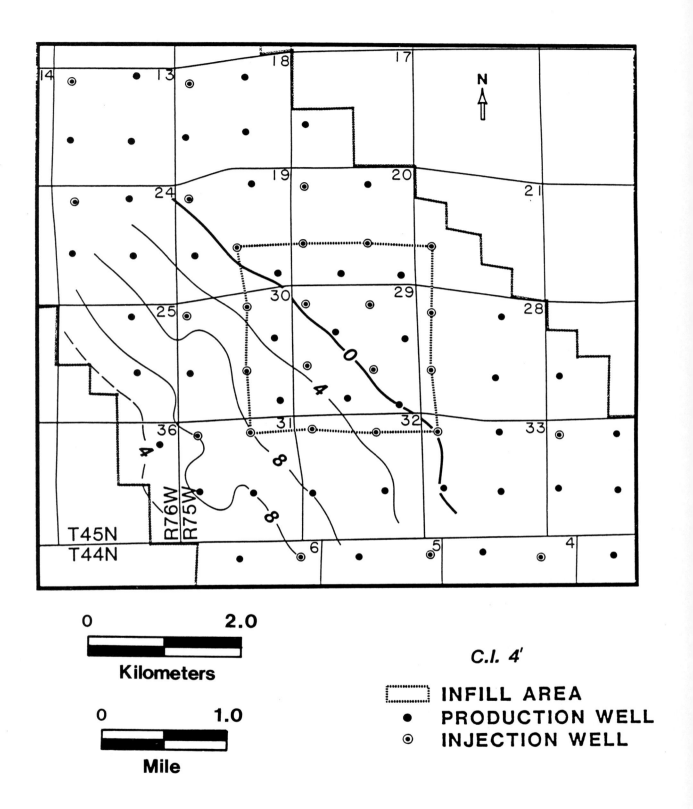

Figure 14. Thickness of the upper lens of the Shannon Sandstone within and around the study area. Contour interval is in feet.

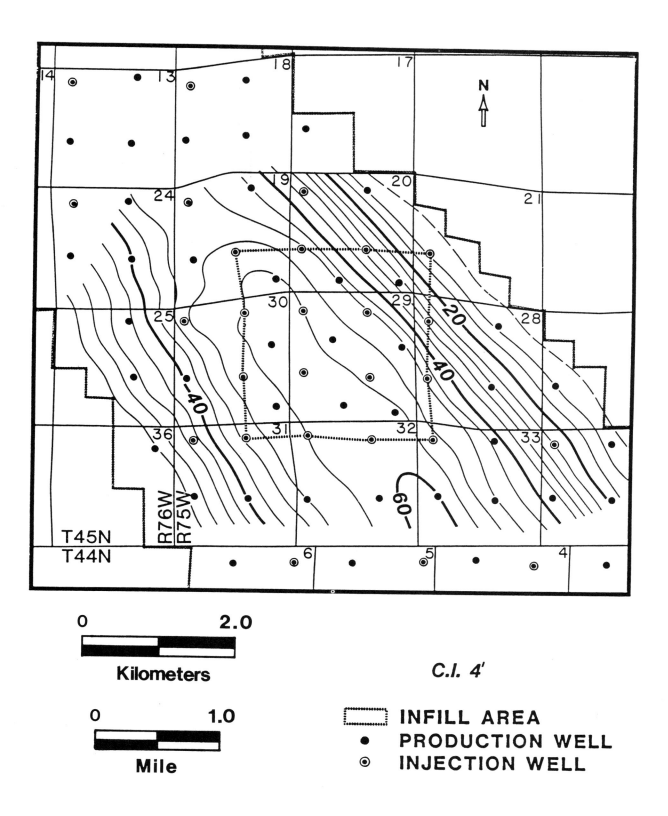

Figure 15. Thickness of the Shannon Sandstone within and around the study area exclusive of the basal facies D section. Contour interval is in feet.

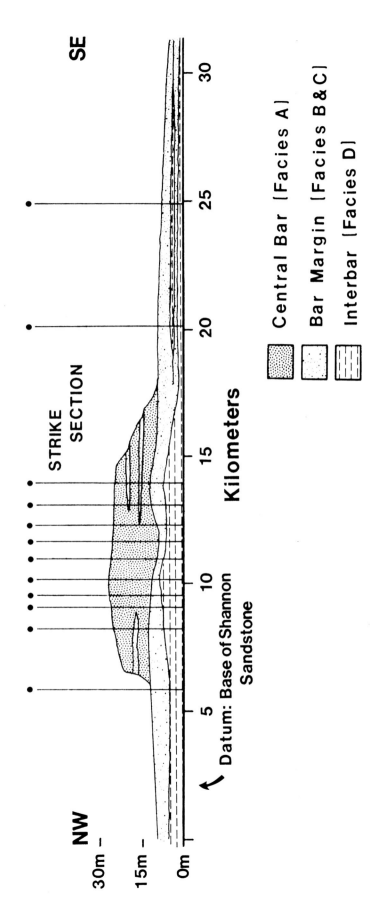

Figure 16. Strike-oriented stratigraphic cross section of the Shannon Sandstone within Hartzog Draw Field. Basal bar-margin (facies B and C) sandstone overlies the interbar (facies D) and has a great lateral extent. Thickest sandstone occurs in the north-central portion of the field and consists of well-developed central-bar (facies A) sandstone interbedded with bar-margin (facies B and C) deposits. Note difference in scale from Figure 17.

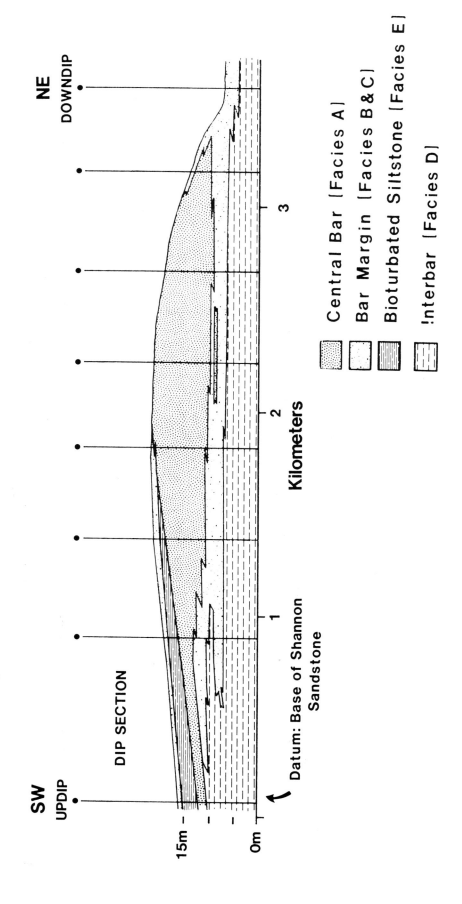

Figure 17. Dip-oriented stratigraphic cross section of the Shannon Sandstone in Hartzog Draw Field. Note the interfingering of central-bar (facies A) with bar-margin (facies B and C) facies, and the updip (SW) division of the sandstone lenses by the bioturbated siltstone (facies E) and interbar (facies D). Note difference in scale from Figure 16.

204

Central-bar and bar-margin deposits interfinger and thin updip into the interbar and bioturbated siltstone facies. This implies the importance of periodic landward sediment transport, possibly by storm waves and currents. Occasionally, thin sequences of central-bar sandstone were deposited in the lower bar-margin facies by unidirectional shelf currents. Such sequences may represent periods of lower sedimentation rates or increased current activity (which formed the laterally discontinuous and thin lenses of the better-sorted central-bar sandstone).

Petrography and Sandstone Diagenesis

Most samples of the Shannon Sandstone in Hartzog Draw Field plot as litharenites and feldspathic sublitharenites, according to the classification of Folk (1968). The average framework composition is 60% quartz, 15% feldspar, and 25% lithic rock fragments. Commonly occurring rock fragments include chert, granite, gneiss, shale, and glauconite. Sandstone composition, the inclusion of clay-size particles, and the presence of carbonate, quartz, and clay cements significantly influence reservoir quality.

When other parameters remain constant, sandstones with more rock fragments, chert, and glauconite have undergone a considerable decrease in primary porosity because of compaction and plastic deformation of grains. Glauconite is often the most ductile constituent and, where abundant, may form a semicontinuous pseudomatrix. Facies A and B sandstones from wells 5202 and 6334 (outside the infill area) have low average thin-section porosities, 3.0% and 2.5%, respectively, because of their higher ductile grain content. Exceptions are sandstones exhibiting early carbonate cementation that prevented compaction. In these sandstones, the cements were subsequently dissolved to form secondary porosity.

Detrital clay content predictably varies with the sedimentary structures present in the sandstone. Higher-energy facies (facies A and B) have very little or no dispersed detrital clay. Clay can occur as mud chips or as thin, discontinuous laminae in ripple-bedded sandstone (facies C and D). Bioturbated sandstone and siltstone (facies D and E) contain the greatest amount of dispersed detrital clay. Porosity and permeability are partly controlled by the amount of dispersed detrital clay. Porosity observed in thin sections decreases from an average of 7.6% in facies A and B, and approaches 0% in facies E owing to an increase in rippling, shale laminations, burrowing, and clay content.

Quartz, carbonate (ferroan calcite and ferroan dolomite), and clay (chlorite and illite/smectite) account for the major cements found in the Hartzog Draw reservoir. Quartz cement content is low (average, 1.6%) in thin sections but appears to be much more abundant under the SEM. Overgrowths are easily seen when outlined by brown, authigenic grain-coating clay, and overgrowth content may be as high as 10%. Where grain coatings are absent, the percentage of quartz cement is underestimated because the overgrowths coalesce to give the appearance of straight "contacts." Carbonate cements are sparse on the average (average, 2.6%), probably because of a large amount of dissolution. Carbonate replacement of framework grains occurs commonly, with feldspars being the most susceptible. The occurrence of dolomite (determined from X-ray diffraction) in mud-rich sandstones may be related to the release of Mg or Fe ions during alteration of smectite to illite in the original matrix.

Siderite cement is very low (average less than 1%). It is sucrosic, microgranular, and brownish in color. Cross-bedded sandstones having 30%-40% siderite cement are present, but these form no more than 5% of the net sand thickness. Siderite cement is almost always restricted to intervals with an abundance of siderite clasts. The cement probably was derived from solution of the siderite clasts and in situ reprecipitation. Such dissolution has formed intra-cement micropores.

Chlorite content estimated from thin sections is low (average, 1.9%), but authigenic chlorite is the most abundant clay cement visible under the SEM. Pore-filling as well as grain-coating iron-rich chlorite occurs as blades 0.005-0.010 mm in diameter. Glauconite grains contain illite or interlayered illite/smectite in the center but are commonly altered to, or coated by, chlorite at their margins. The amount of chlorite less than 0.005 mm measured by XRD increases with the amount of glauconite determined by point-counting. This suggests that some of the chlorite may have been locally derived from the alteration of glauconite grains.

Pore-filling authigenic illite and interlayered illite/smectite occur in moderate amounts in a few samples but are sparse in most. Authigenic illite is recognizable by its "cornflake" morphology and by the presence of a potassium peak in the X-ray spectrum. Authigenic K-feldspar and albite cements are abundant in a few samples but sparse overall. Minor amounts of authigenic framboidal pyrite and dispersed dodecahedral crystals of pyrite occur in many samples.

The following paragenetic sequence summarizes the diagenetic events that modified the sandstone framework and pore space: (1) siderite cementation; (2) precipitation of brown, grain-coating clay; (3) a first stage of quartz precipitation; (4) precipitation of calcite and dolomite; (5) dissolution of carbonate cements and some framework grains; (6) precipitation of quartz cement and additional grain-coating clay, chlorite, and possibly some illite/smectite; and (7) oil emplacement. For a further explanation of the diagenesis, see Ranganathan and Tye (1986).

Flow Units

This study was begun with the hope that a better understanding of the geology of this Shannon Sandstone reservoir would improve predictions of primary oil recovery and enhance reservoir simulation models to predict secondary and tertiary recovery. Considerable variability was found in the reservoir properties of each "facies"; therefore, a somewhat different subdivision of the reservoir into five flow units was necessary. These flow units are defined and determined by the distribution of the rock characteristics that most strongly control reservoir behavior as fluids are injected or produced (Hearn et al., 1983).

The stratigraphic sequence of facies is the framework used to define the flow units, but the flow units do not coincide exactly with facies. Flow-unit divisions are based not only on geologic characteristics and position in the vertical sequence of the geologic facies, but also on their petrophysical properties, especially porosity and permeability. No unique value of these properties defines a flow unit; rather, certain ranges of porosity and permeability that occur in a particular part of the sedimentary sequence are used to subdivide the reservoir along lines that represent gradations in reservoir quality (i.e., the ability of the rocks to transmit fluids, both laterally and vertically).

This procedure of designating and correlating flow units is somewhat subjective, but when numerous cross sections are constructed within an area of interest, consistent patterns of continuity of flow units can be detected (Fig. 18). The value of this approach to subdivision of the reservoir is twofold. It allows for more quantitative definition and mapping of the parts of the sandstone sequence that are most important to reservoir behavior, and it forms a fairly realistic basis for definition of reservoir zonation for numerical simulation of reservoir performance. In a large field such as Hartzog Draw, in which considerable variability of rock properties exists both vertically and laterally, the delineation of flow units to be used in numerical reservoir models is most important. For example, flow unit 4 (Fig. 19) represents the most homogeneous, most consistently porous and permeable part of the central bar (facies A) and is apparent only by division of this facies. A comparison of the isopach of flow unit 4 (Fig. 19) with that of the middle sandstone lens (Fig. 13) reveals the rapid westward degradation of good-quality reservoir sandstone. Although the middle lens retains a thickness of nearly 4 m near the westward margin of the field, flow unit 4 terminates just westward of the northwest-southeast axis of the field. This comparison does indicate, however, that flow unit 4 has good continuity along strike (Fig. 19).

That the flow units are representative of fairly distinctive rock characteristics is further indicated by the relationship of the average

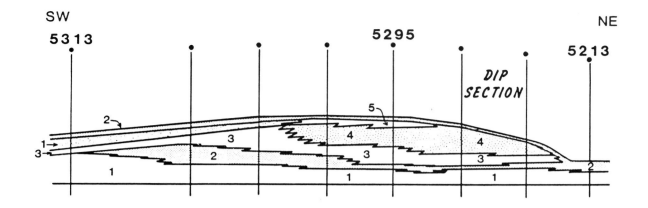

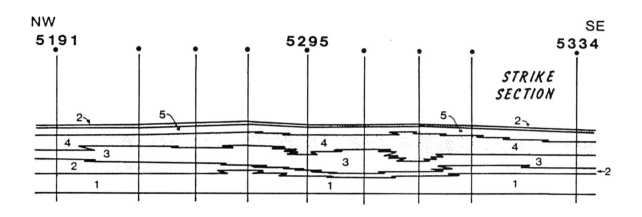

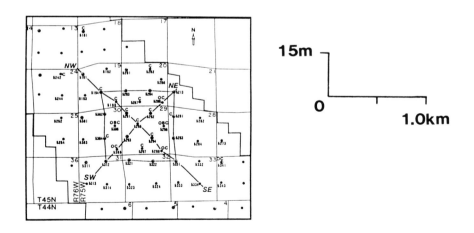

Figure 18. Stratigraphic cross sections illustrate the arrangement of flow units within the Shannon Sandstone in the study area. Note asymmetry of flow-unit distribution in dip (SW-NE) section and relationship to isopach of Shannon (Fig. 15) and to cumulative oil production pattern (Fig. 3). See Figure 2 for locations.

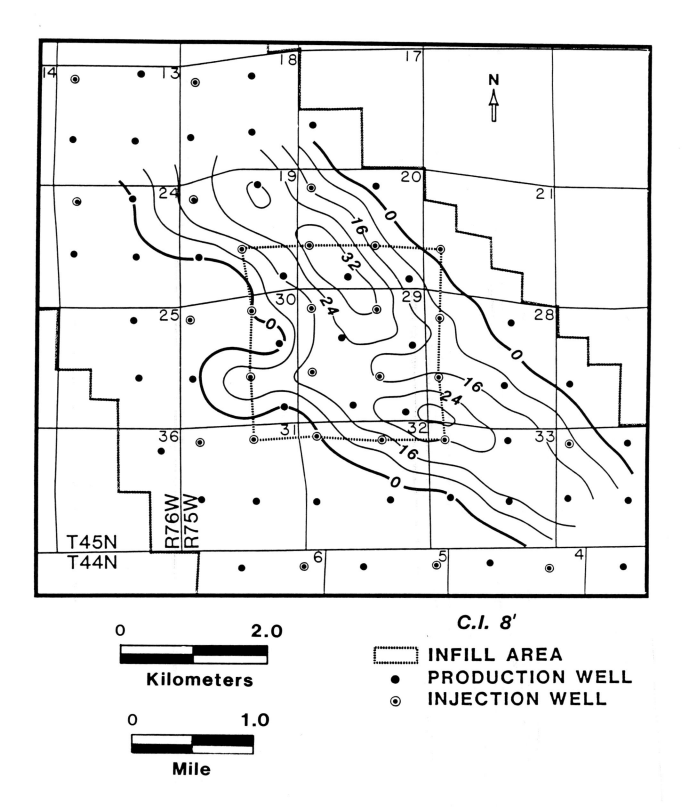

Figure 19. Thickness of flow unit 4 in the study area. Flow unit 4, where thickest, contains the most permeable and porous parts of facies A and has produced the largest volume of oil per porosity foot.

porosity and permeability values of each flow unit to the whole population of measurements made on cores from Hartzog Draw. Furthermore, pore-size distributions were determined for numerous samples from mercury injection data. Although there was considerable scatter in the data, average pore-size distribution curves for each flow unit suggest differences in the nature of the pore networks. Average properties of each of the flow units vary areally, corresponding to gradational facies changes and to gradients in diagenetic modifications of the pore systems.

Cross sections of flow-unit distribution in the pilot area (Fig. 18) bear some resemblance to the stratigraphic cross sections shown in Figures 16 and 17. Because flow units are based on depositional facies, petrology and diagenesis, and petrophysical data, the most homogeneous flow units (units 4 and 5) only partially mimic the distributional patterns of Facies A and B.

The subdivision of the reservoir into flow units aids in understanding present reservoir performance and predicting future oil recovery. Areally, primary production performance follows the trends of the flow units. For example, flow unit 4 dominates the east side of the reservoir in the pilot area (Fig. 19) and has the highest permeability and porosity (Fig. 5). Comparison of Figure 19 with Figure 3 shows that the areas of greatest thickness of flow unit 4 correspond fairly well with the areas of greatest cumulative primary production.

Conclusions

The Shannon Sandstone in the Hartzog Draw Field, Campbell County, Wyoming, was deposited as a lenticular shelf sand ridge on the western margin of the Cretaceous Seaway. The following six lithofacies are recognizable in cores and are ranked in order of decreasing porosity and permeability: (1) facies A (medium-grained, cross-bedded sandstone), (2) facies B (medium- to coarse-grained, cross-bedded sandstone), (3) facies C (fine-grained, laminated, rippled sandstone), (4) facies D (fine-grained, burrowed, shaly sandstone), (5) facies E (bioturbated siltstone), and (6) facies F (shale). The five sandstone and siltstone lithofacies compose a lenticular, upward-coarsening, and elongate shelf sand ridge encased in shale. Deposition and preservation of this sand ridge created a stratigraphic trap.

The reservoir consists of three stacked sandstone lenses (upper, middle, and lower) separated by siltstone and shale intervals. Shaly sandstones of facies B and C comprise most of the upper and lower lenses, thus accounting for their relatively poor reservoir quality, whereas the thicker middle lens consists almost entirely of clean Facies A sandstone interbedded with thin facies B and C deposits. Analyses of wire-line logs, cores, thin sections, and core porosity and permeability data indicate that reservoir performance is influenced by (1) sandstone stratification and internal sedimentary structures, (2) the occurrence and extent of shale laminations, and (3) cementation.

Flow units defined within the stratigraphic and sedimentologic framework of the reservoir permit delineation of the homogeneities that most strongly influence reservoir fluid behavior. Flow units are defined on the basis of variations in permeability, well-log character, geologic character, and stratigraphic occurrence. The recognition of flow units delineates and improves the prediction of porous and permeable zones within the reservoir and has proved more valuable in reservoir simulation that sandstone isopach maps and geologic cross sections based solely on lithofacies.

Acknowledgments

Thanks are extended to Cities Service Oil Company, Tulsa, Oklahoma, for permission to publish this information. This study benefited from numerous discussions with J. M. Boyles, C. L. Hearn, T. F. Moslow, and R. W. Tillman. Jane Lemoine typed numerous versions of this manuscript.

References

Berg, R. R., 1975, Depositional environment of the Upper Cretaceous Sussex Sandstone, House Creek Field, Wyoming: Bull. Amer. Assoc. Petrol. Geol., v. 59, p. 2099-110.

Berven, R. J., 1966, Cardium sandstone bodies, Cross Garrington Area, Alberta: Can. Jour. Petrol. Geol., v. 14, p. 208-40.

Boyles, J. M., and Scott, A. J., 1982, A model for migrating shelf bar sandstones in Upper Mancos Shale (Campanian), northwestern Colorado: Amer. Assoc. Petrol. Geol., Bull., v. 66, p. 491-508.

Evans, W. E., 1970, Imbricate linear sandstone bodies of Viking Formation in Dodoland Hossier area of southwestern Saskatchewan, Canada: Amer. Assoc. Petrol. Geol. Bull., v. 54, p. 469-86.

Folk, R. L., 1968, Petrology of Sedimentary Rocks: Austin, Tex., Hemphill Publishing Co., 182 p.

Gill, J. R., and Cobban, W. A., 1969, Paleogeographic map for latest Eagle time showing position of strandplain during range zone of _Baculites_ sp. (smooth) = U.S. Geological Survey Open File Report 1969, sheet 2 of 6 paleogeographic maps.

Hearn, C. L., Ebanks, W. J., Tye, R. S., and Ranganathan, V., 1983, Geologic factors influencing reservoir performance of the Hartzog Draw Field, Wyoming: Jour. Petrol. Tech., Aug. 1984, p. 1335-44.

Hobson, J. P., Fowler, M. L., and Beaumont, E. A., 1982, Depositional and statistical exploration models, Upper Cretaceous offshore sandstone complex, Sussex Member, House Creek Field, Wyoming: Amer. Assoc. Petrol. Geol. Bull., v. 66, p. 689-705.

Hunt, R. D., and Hearn, C. L. 1981, Reservoir management of the Hartzog Draw Field: Soc. Petrol. Engin., Paper SPE 10195, 6 p.

Martinsen, R. S. and Tillman, R. W., 1979, Facies and reservoir characteristics of a shelf sandstone, Hartzog Draw Field, Powder River Basin, Wyoming (Abs.): Amer. Assoc. Petrol. Geol. Bull., v. 63, p. 491.

Porter, K. W., 1976, Marine shelf model, Hygiene Member of the Pierre Shale, Upper Cretaceous, Denver Basin, Colorado, in Epis, R. C. and Wiemer, R. J., eds., Studies in Colorado field geology: Colorado School of Mines, Professional Contribution No. 8, p. 251-63.

Ranganathan, V., and Tye, R. S., 1986, Petrography, diagenesis and facies controls on porosity in the Shannon Sandstone, Hartzog Draw Field, Wyoming: Amer. Assoc. Petrol. Geol. Bull., v. 70, p. 56-69.

Tillman, R. W., and Martinsen, R. S., 1979, Hartzog Draw Field, Powder River Basin, Wyoming, in Flory, R. W., ed., Rocky Mountain high: Wyoming Geol. Assoc., 28th Ann. Meeting, Core Seminar Book, p. 1-38.

Tillman, R. W., and Martinsen, R. S., 1984, The Shannon Shelf-Ridge Complex, Salt Creek Anticline Area, Powder River Basin, Wyoming, in Tillman, R. W., and Siemers, C. T., eds., Siliciclastic shelf sediments: Soc. Econ. Pal. Min., Spec. Pub. No. 34, p. 85-142.

CARDIUM FORMATION CONGLOMERATES AT CARROT CREEK FIELD:

OFFSHORE LINEAR RIDGES OR SHOREFACE DEPOSITS?

Katherine M. Bergman and Roger G. Walker

Department of Geology, McMaster University
Hamilton, Ontario, Canada L8S 4M1

Abstract

Carrot Creek field consists of a series of long, narrow conglomerate bodies oriented roughly parallel to regional strike and encased in marine shales. Here, the Upper Cretaceous Cardium Formation comprises two coarsening-upward sequences. The lower "b" sequence contains offshore bioturbated mudstones and is capped by a gritty siderite interpreted to represent a pause in deposition. The upper "a" sequence begins with bioturbated mudstones, overlain by hummocky cross-stratified sandstones and conglomerates. Sandstone-conglomerate bodies of this type have traditionally been interpreted as offshore ridges or "terrace bars," posing major problems concerning sand and gravel transport from a distant (unidentified) shoreline.

Our cross sections clearly demonstrate a major erosional surface, of about 20 m relief, at the base of the conglomerate. The conglomerate can rest on various facies within the "b" and "a" sequences but is not genetically related to them. A structure map of the erosional surface shows three topographically distinct areas: (1) a rather smooth terrace, (2) a major bevel where the "b" and "a" sequences are truncated, and (3) a basinward erosional topography of remnant bumps and hollows.

The erosion is believed to have taken place during a rapid relative lowering of sea level, and the erosional bevel represents a newly established shoreface. Gravel supplied at lowstand was reworked along this shoreface and into the hollows by waves. The elongate gravel accumulations were buried by marine shales during the ensuing transgression.

Introduction: Questions Posed by Linear Ridges

This is the ninth in a series of papers discussing aspects of the sedimentology of the Upper Cretaceous Cardium Formation of Alberta, Canada. Many formations in the Western Interior Seaway, including the Cardium, contain a series of linear sandstone and conglomerate bodies. These have traditionally been regarded as long, narrow, en echelon ridges or bars, parallel or subparallel to regional strandlines and completely encased in marine shales. They will be termed "ridges" in the following text.

In many of the ridges, the sedimentary sequence coarsens upward, suggesting that the sandstones and conglomerates at the tops of the ridges are gradationally rooted in offshore marine shales. The origin of the ridges is problematic because of their apparent depositional setting, commonly many tens of kilometers from the nearest shoreline.

The Cardium Formation contains several such ridges, and findings concerning their origin might be applicable to similar ridges in other formations. The conglomeratic nature of the linear ridges in the Cardium makes understanding their development particularly difficult.

In the Carrot Creek oil field (Fig. 1), the en echelon ridges are up to 19 m thick and have no connection to any known shoreline (Plint et al., 1986). If the conglomerate ridges were indeed deposited far from the shoreline, as suggested by Swagor et al. (1976), they present two substantial problems:

1. How was the gravel transported across the shelf?
2. How was the gravel focused or molded into long, narrow en echelon ridges?

Various combinations of storm-generated geostrophic currents, density currents, and tidal currents may be capable of transporting sufficient volumes of sand and gravel seaward across the shelf (as reviewed by Swift and Niedoroda, 1985). No one has convincingly explained how these currents focus or mold the coarse sediment into long, narrow ridges. The traditional interpretation does not explain why these ridges contain coarsening-upward sequences in which gravel is selectively transported to the tops of the ridges. These problems

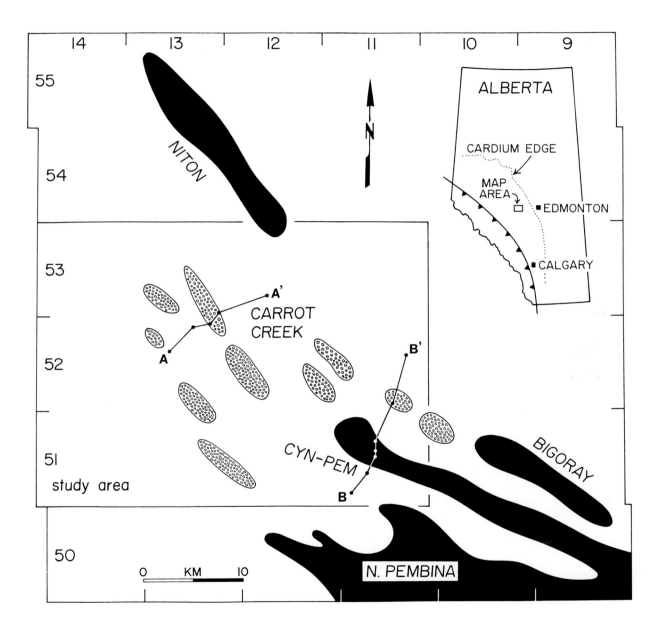

Figure 1. Location map showing Carrot Creek conglomerate pods (circles) and adjacent Cardium fields (black). Note location of cross section A (Fig. 4) in northwest and cross section B (Fig. 5) in southeast, crossing both Cyn-Pem and Carrot Creek fields. Inset shows the location within Alberta; the thrust symbol denotes the edge of the deformed belt of the foothills, and the dotted line shows the northeastern extent of Cardium sand.

have forced us to reevaluate whether the bodies are in fact ridges and whether they were actually deposited many kilometers seaward of the shoreline.

Origin of the Carrot Creek Conglomerates

Existing discussions of Cardium sedimentation emphasize offshore environments, with storm-influenced deposition below fair-weather wave base but above storm-wave base (Swagor et al., 1976; Walker, 1983a; Krause and Nelson, 1984; Keith, 1985). For the Carrot Creek conglomerates, Swagor et al. (1976) specifically suggested storm transport of sediment across the shelf and deposition in "terrace bars" in the lee of preexisting topographic features. They implied that the sediment moved as bed load. To account for features in conglomerates elsewhere in the Cardium, and to account for an increase in rate of gravel emplacement, Wright and Walker (1981) suggested that storm-generated turbidity currents transported gravel across the shelf.

Our data from Carrot Creek suggests that we may have been asking the wrong questions. In the subsurface (Plint et al., 1986) and in outcrop (Duke, 1985), the Cardium is characterized by a series of regionally extensive erosion surfaces, numbered (in subsurface) E1-E7 in Figure 2. At least a veneer of conglomerate covers each surface, and at Carrot Creek, up to 19 m of conglomerate covers surface E5. The way the surfaces truncate well log markers and the absence of various facies recognized in core demonstrate that the surfaces are erosional. The erosion presumably resulted from wave scouring of the bed, and hence the erosion surfaces probably represent periods of relative lowering of sea level (Plint et al., 1986). During such periods, the shoreline moved rapidly basinward, and newly incised rivers supplied

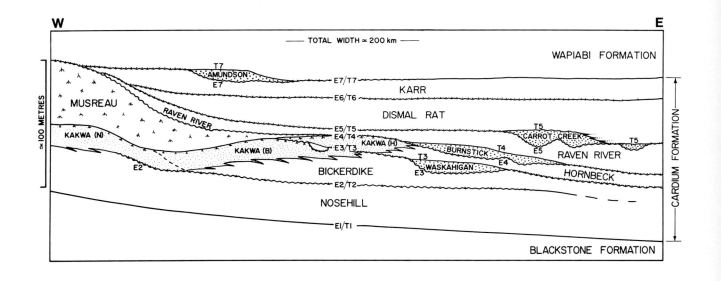

Figure 2. Member names for the Cardium Formation in the subsurface (from Plint et al., 1986). The event stratigraphy is defined by a series of erosion (E) and transgressive (T) surfaces, numbered in the figure.

coarse material to the new shoreface. During the ensuing transgression, some of the gravels will be at first reworked as lags, but as the water deepens, the area will be blanketed by basinal muds.

An alternative possibility is that the erosion surface was cut subaerially. However, the conglomerate bodies are oriented parallel or subparallel to regional strike (and by implication, to regional shorelines). If these conglomerates were fluvial, their regional orientation would be normal to strike. The orientation of all of the bodies is very consistent (NW-SE); if they were fluvial (channels and/or bars), much more variability in strike would be expected. This study demonstrates that the orientation of the conglomerate bodies is related to the morphology of the erosion surface. The one-sided geometry of the bevel (with its erosional edge on the landward side and open on the seaward side), and the absence of any evidence of channelization, suggest that the erosion is <u>not</u> the result of fluvial downcutting during lowstand.

We suggest that Carrot Creek represents a major drop in relative sea level, and the erosional topography is due to waves scouring a new shoreface. This interpretation is consistent with the preservation of coarse material completely encased in marine mudstones, and it explains both the transportation of coarse material across the shelf and its deposition in long, narrow ridges. Once the coarse gravel is delivered to the new shoreface, the longshore drift system can rework it into elongate pods and preferentially deposit it in the low areas of the erosional topography.

Although the effects of major transgressions and regressions are becoming reasonably well understood in the Western Interior Seaway

(Kauffman, 1969, 1977; Caldwell, 1984; Weimer, 1984), little detailed work exists on the effects of smaller-scale (and perhaps more rapid) fluctuations. The seven Cardium erosion surfaces identified by Plint et al. (1986) represent seven important erosional lowerings of relative sea level, probably reflecting tectonic (Jeletzky, 1978), as well as eustatic, controls. These relative sea level fluctuations are superimposed upon the one global eustatic regression in the late Turonian that marks the end of Kauffman's (1969, 1977) Greenhorn Cycle, termed R_6.

Our results concerning the Cardium Formation suggest that other offshore ridges should be reexamined in terms of rapid and subtle sea level changes that result in rapid shoreline movements, rather than in terms of shelf sediment-transport processes. Recent work on the Upper Cretaceous Bad Heart Formation (A. G. Plint, pers. comm., 1986) has documented a surface of about 50 m relief scoured into hummocky cross-stratified sandstones and mudstones and covered with marine mudstones. The Lower Cretaceous Viking Formation at the Gilby (Raddysh, 1986) and Joffre fields (K. P. Downing, pers. comm., 1986) consists of coarse- and very coarse-grained sandstones in long, narrow bodies. Two erosion surfaces have recently been documented beneath the Joffre field (K. P. Downing, pers. comm., 1986), suggesting deposition as a newly incised shoreface sand body rather than an "offshore ridge" deposited an unknown distance from the shoreline.

It follows that coarsening-upward sequences that end with coarse sandstone or conglomerate in the Western Interior Seaway are not necessarily single genetic packages. Many of these sequences are composite; the lower parts reflect aggradation on a marine shelf, and the upper parts reflect a relative sea level drop, incision of a new shoreface, and deposition of coarse sand and gravel on an erosion surface.

Regional Setting

The Upper Cretaceous (Turonian) Cardium Formation comprises mudstones, sandstones, and conglomerates. It is encased in 750 m of marine shales of the Alberta Group (Colorado Group), 250 m of Blackstone Formation shales below and 500 m of Wapiabi Formation shales above (Fig. 3). The Cardium is an approximate time equivalent of the Frontier, Ferron, and Gallup sandstones located in the central and southern portions of the Western Interior Seaway.

The Cardium Formation is about 100 m thick and crops out in the foothills of the Canadian Rockies, where it contains a stacked series of coarsening-upward sequences. Similar coarsening-upward sequences also occur in the subsurface of the Alberta Plains (Fig. 1), but to date, no satisfactory correlation has been found between the two areas.

Until recently, there has been no formal stratigraphic subdivision of the Cardium Formation in the subsurface. Plint et al. (1986) recognized and correlated a series of erosion surfaces, E1-E7 (Fig. 2), overlain by conglomerates varying from a thin veneer to about 20 m thick. The conglomerates are overlain by fine-grained mudstones; the contact is taken to be the transgressive surface. The transgressive surfaces are numbered T1-T7 (Fig. 2). The E and T surfaces are essentially coincident where the conglomerate is reduced to a thin veneer and are referred to, for example, as E4/T4. Plint et al. (1986) now formally draw the base of the Cardium Formation at the E1/T1 surface and the top at the E7/T7 surface (Plint et al., 1986).

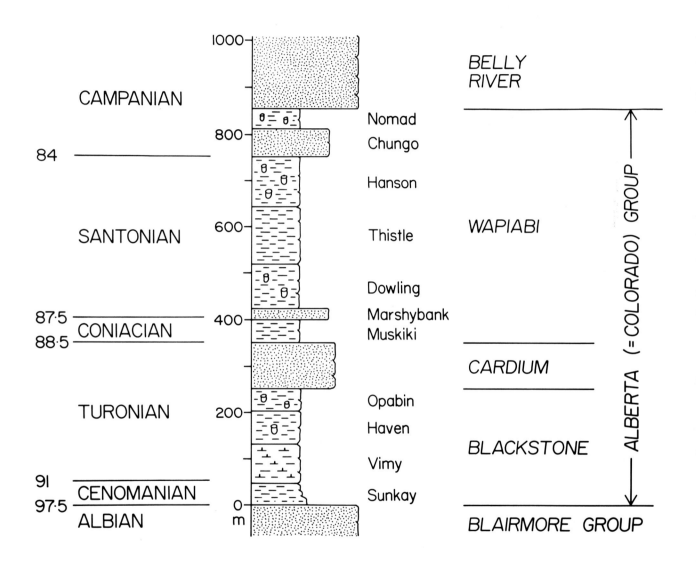

Figure 3. Stratigraphy of the Alberta (Colorado) Group in the west-central Alberta Basin, with absolute ages (in millions of years) shown at left.

In the Carrot Creek study area, the erosional and transgressive surfaces shown in Figure 2 define three members. The Raven River Member includes all of the coarsening-upward sequences between T4 and E5. The Carrot Creek Member includes the thick conglomerates but not the mudstones with pebbly stringers between E5 and T5. The Dismal Rat Member (named after the confluence of Dismal Creek and Rat Creek) includes the mudstones that overlie T5 (including facies 3P, 4P, and 5P), up to the log marker known to industry as the Cardium "Zone" (E6/T6).

Carrot Creek Oil Field

Carrot Creek is approximately 288 km north-northwest of Calgary (Fig. 1), located primarily in townships 52 and 53, ranges 11-14 W5. The field was discovered in 1963; the discovery well was Pan-Am A-1 Carrot Creek (16-7-52-12W5). The estimated in-place oil reserves are 1124.4×10^3 m^3, or 7.07 millions barrels (additional reserves from water flooding of Energy Resources Conservation Board Cardium A and F pools are 322×10^3 m^3, or 2.02 million barrels). The sandstones are tight. Oil production is from the conglomerates, which are up to 19 m thick. Clast-supported facies have porosities in the 6%-12% range, with permeabilities ranging from about 1-100 md.

The data base consists of 270 well logs; of these wells, 135 are cored. The study is based on gamma-ray and resistivity-log signatures. The gamma-ray trace tends to mirror the resistivity trace and is of limited availability; consequently, only the resistivity trace is shown. Log and core cross sections are shown in Figures 4 and 5 and are located in Figures 1, 6, and 7. Log markers above and below the conglomerates were easily recognized and correlated, particularly the

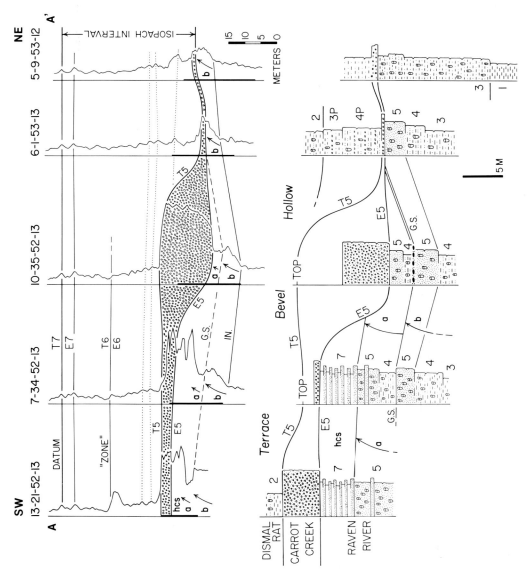

Figure 4. Cross section A-A', located in Figures 1, 6, and 7. Cores (below) are hung on the same E7/T7 marker as the resistivity logs (above). The cored interval in each well is indicated by a solid bar. Numbers indicate facies described in the text; the vertical cylindrical symbol indicates bioturbation. "G.S." indicates gritty siderite, and "IN" marks inflection points in resistivity logs. Arrows indicate coarsening-upward sequences "a" and "b." The horizontal separation between wells does not represent the true distance.

227

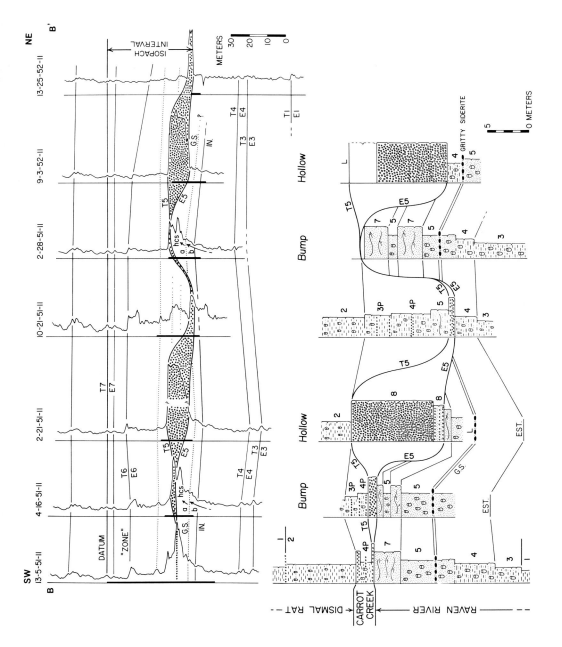

Figure 5. Cross section B-B', located in Figures 1, 6, and 7. Symbols and numbers are the same as in Figure 4; note also the hummocky cross-stratification symbol in facies 7. "EST" indicates the estimated position of the facies 3/4 boundary. The stratigraphic sequence within the topographic bumps (13-5, 4-16, and 2-28) is identical to that within the terrace (Fig. 4, 13-21, 7-34). The horizontal separation between wells does not represent the true distance.

228

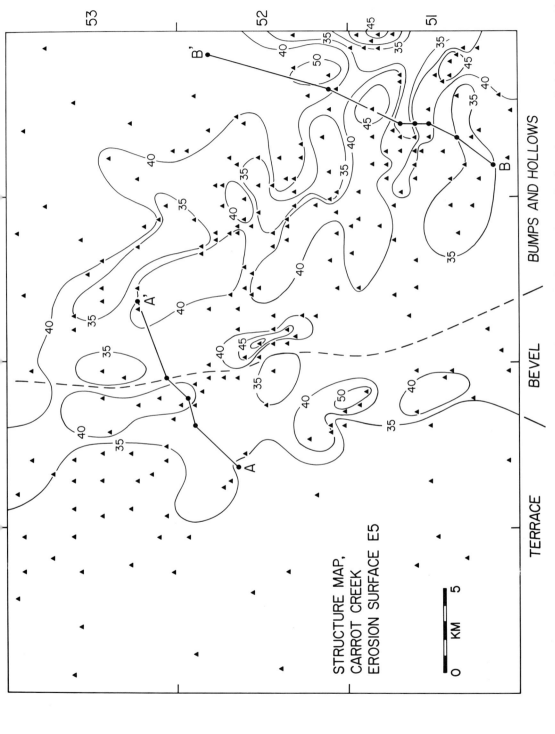

Figure 6. Structure map of the erosion surface E5 showing isopachs of the interval between the E7/T7 marker and the base of the conglomerate (E5) in meters. Cored wells are shown with solid triangles. As labeled at the base of the figure, note terrace, generally less than 35 m below marker; bevel (outlined on map by solid and dashed lines), where the E5 erosion surface drops basinward; and bumps and hollows, an erosional remnant topography.

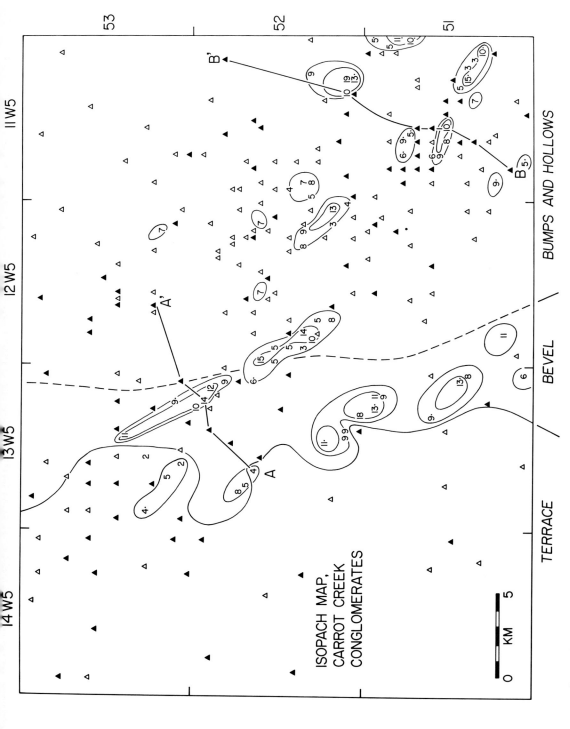

Figure 7. Conglomerate isopach map. Cored wells are shown with solid triangles; open triangles indicate well logs only. All wells show at least a veneer of conglomerate. Where the thickness exceeds 2 m, the well location is shown by a number indicating the actual thickness. A period after the number indicates no core control. Isopach lines show thicknesses greater than 2 m, or greater than 10 m.

upper datum (E7/T7), the lower markers E3/T3 and E4/T4, and the prominent inflection point (IN) in the induction log. A distinctive, gritty siderite layer in the core gives the subtle log response labeled "G.S." (Figs. 4, 5). The base of the conglomerate forms an *erosion surface*, E5.

The cored interval in each well is indicated by a solid bar. Facies and facies sequences were measured and defined in all 135 cores and were correlated with the well logs by picking an obvious core and log marker (e.g., top of the conglomerate). In this way the core depth is adjusted to the log depth. The separation between the wells in the cross sections does not represent the true distance between wells.

Core and Facies Descriptions

Three cores typifying the facies sequences in the Carrot Creek area were chosen to illustrate the three major areas suggested by the geometry of the erosion surface E5 (Fig. 6). Most of the facies found in the Carrot Creek area are essentially the same as those described by Walker (1983B) for the Raven River Member at Caroline and Garrington fields farther south, so only a brief description of the facies is presented.

Facies 1, Massive Dark Mudstone

Facies 1 (Fig. 8A) is found at the base of the Raven River Member and above the "laminated blanket" of the Dismal Rat Member (Fig. 4). These are structureless mudstones containing no recognizable burrow forms, although there is a faint mottling (probably made by the pinworm *Gordia*).

Figure 8. [A] Massive dark mudstones of facies 1; 5597 ft (1749.1 m) in well 13-5-51-11. This detail photo is from above the interval shown in the boxed core of Figure 13. It is from the facies 1 occurrence at the top of the measured core in Figure 5. A 3.0-cm scale is shown in most close-up core photographs. [B] Laminated dark mudstones of facies 2; 5630.5 ft (1759.5 m) in well 13-5-51-11 (see Fig. 13). Scale in centimeters.

Facies 2, Laminated Dark Mudstones

Facies 2 (Fig. 8B) mudstones comprise the "laminated blanket" of the Dismal Rat Member (Fig. 4). These mudstones differ from those of facies 1 by the preservation of thin, silty laminae less than 1 cm thick. There are no recognizable burrow forms, although the background mudstones are faintly mottled, and the laminae may be somewhat disrupted.

Facies 3, Dark Bioturbated Muddy Siltstones

Facies 3 (Fig. 8C) gradationally overlies facies 1 and contains remnants of silty laminae, indicating extensive bioturbation. There are no distinct burrow forms.

Facies 4, Pervasively Bioturbated Muddy Siltstones

Facies 4 (Fig. 8D) is a pervasively bioturbated siltstone that has some preserved sharp-based, wave-rippled, graded beds (1-5 cm thick). Recognizable traces include *Rhizocorallium, Teichichnus,* and *Terebellina.*

Facies 5, Bioturbated Sandstone

Facies 5 (Fig. 9A) is similar to facies 4 but contains more sand. Silty, sharp-based, wave-rippled, graded beds up to 7 or 8 cm thick may be preserved. Bioturbation is intense and recognizable traces include *Rhizocorallium, Zoophycos, Teichichnus, Terebellina, Skolithos,* and *Chondrites.*

Figure 8. [C] Dark, bioturbated muddy siltstones of facies 3; 5698 ft (1780.6 m) in well 13-5-51-11 (see Fig. 13). Scale in centimeters. [D] Pervasively bioturbated muddy siltstones of facies 4; 5687.5 ft (1777.3 m) in well 13-5-51-11 (see Fig. 13). Note *Teichichnus* burrow (arrow) in the center of the photograph. Scale in centimeters.

Figure 9.   [A] Bioturbated sandstones of facies 5; 5664.5 ft (1770.2
m) in well 13-5-51-11 (see Fig. 13).   Note the abundance of
Chondrites burrows (arrow), a feature characteristic of this
facies.   Scale   in   centimeters.   [B]   Nonbioturbated
sandstones of facies 7; 5647 ft (1764.7 m) in well 13-5-51-11
(see Fig. 13).   Laminae immediately above the scale dip
gently to the left; laminae in the upper part of the photo
are horizontal.   Similar low-angle laminations (interpreted as
hummocky cross-stratification) can be seen in Figure 13,
5658-5645 ft (1768.1 m-1764.1 m).   Scale in centimeters.

Facies 6, Speckled Gritty Mudstone

Facies 6, defined by Walker (1983B), does not occur at Carrot Creek.

Facies 7, Nonbioturbated Sandstone

Facies 7 (Fig. 9B) contains interbedded, sharp-based sandstones and mudstones. The sands are very fine- to fine-grained, and beds display massive to parallel lamination, low-angle inclined stratification (interpreted as hummocky cross-stratification), and wave ripples (which may climb). The sands may be amalgamated and contain mud rip-up clasts (in places sideritized). Sand thickness varies from 10 cm to tens of centimeters. Very little bioturbation is associated with this facies.

One variant of this facies consists of graded sands that are 3–5 cm thick, sharp based, and wave rippled. Thinner beds (1–2 cm) of very black nonbioturbated mudstones separate the sands.

Facies 8, Conglomerates

For the purpose of this paper, the conglomerates can be divided into three types. Thicknesses vary from a thin veneer to a maximum of 19 m.

Clast-supported conglomerates. Clast-supported conglomerates (Figs. 10A-D) form the bulk of the conglomerates in the Carrot Creek area. Clast size varies from less than 1 cm to about 8 cm; average size is 1–2 cm. Stratification is rare, but many textural variations defined by grain size, sorting, and the presence or absence of matrix exist. The lower portion of the conglomerate is bedded (Fig. 10A-C), whereas the upper portion is massive (Fig. 10D).

Figure 10. [A] Clast-supported conglomerate of facies 8; 5426 ft (1695.6 m) in well 2-21-51-11 (see Fig. 14). Note loading of conglomerate into underlying sandstone. Scale in centimeters. [B] Clast-supported conglomerate of Facies 8; 5428 ft (1696.2 m) in well 2-21-51-11 (see Fig. 14). Note wave-reworked sandstone with pebble laminae below and sharp-based conglomerate above. Within the conglomerate, the matrix disappears upward and grain size becomes coarser. Scale in centimeters.

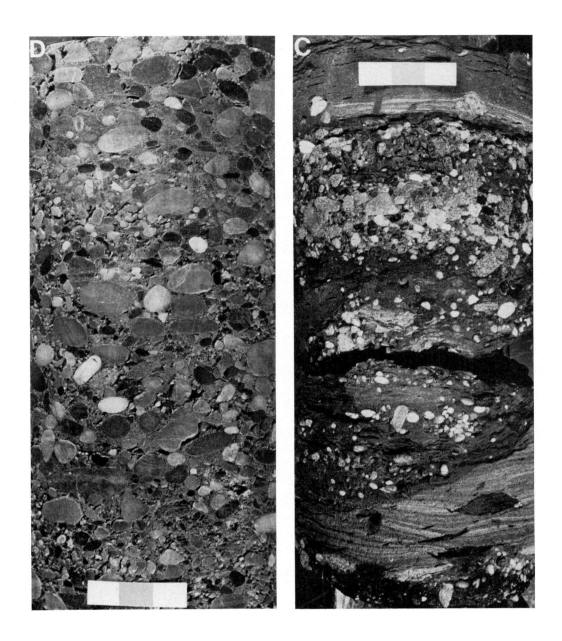

Figure 10. [C] Mud and clast-supported conglomerate of facies 8; 5385 ft (1682.8 m) in well 10-21-51-11 (see Fig. 15). Part of the texture of this pebbly mudstone may be due to bioturbation. Conglomerates of this type occur either at the base of a thick conglomerate sequence or in wells with little conglomerate development (as in 10-21, Fig. 5). Scale in centimeters. [D] Clast-supported conglomerate of facies 8; 5388 ft (1683.8 m) in well 2-21-51-11 (see Fig. 14). This photo is typical of the massive or crudely stratified conglomerates found on the paleo-landward (southwestern) sides of hollows. Scale in centimeters.

Mudstones with conglomerate stringers. The mudstones with conglomerate stringers (Fig. 11A, B) are designated 3P, 4P, and 5P in Figures 4 and 5, indicating that they are similar to facies 3, 4, and 5, but contain scattered chert pebbles and pebble stringers. The stringers vary from one pebble layer thick to a few centimeters thick. Where facies 3P, 4P, and 5P are developed, they always overlie the main clast-supported conglomerates.

Gritty siderite. The gritty siderite facies (Fig. 12A) was not recognized by Walker (1983b) in the Caroline-Garrington area. It is characterized by an abrupt change from very fine-grained quartz sand in the underlying and overlying bioturbated facies to coarser chert grains associated with the gritty siderite. The coarser grains range from medium-grained sand to sand several millimeters in diameter. The siderite distribution is often patchy, with sideritized *Chondrites* burrows. The facies is completely bioturbated, and coarser chert grains occur in the burrows. The thickness varies from 20 to 30 cm over the entire field of study.

Facies Sequences

In the Carrot Creek area the Raven River Member consists of two coarsening-upward sequences (Figs. 4, 5). The lower one ("b" sequence) begins with massive dark mudstones of facies 1 and coarsens upward through facies 3 and 4 into the bioturbated sandstones of facies 5. These are overlain by the gritty siderite horizon (G.S.). The second sequence ("a" sequence) begins with facies 4 or 5 and coarsens upward into the hummocky cross-stratified (HCS) sandstone of facies 7.

Figure 11. [A] Mudstones with conglomerate stringers of facies 3P (see Figs. 4, 5); 5481 ft (1712.8 m) in well 12-16-51-11 (not shown in Fig. 4 or 5). Bioturbated stringers can be seen toward the base and in the center of the core. The width of the core is about 7 cm. [B] Mudstones with conglomerate stringers of facies 4P (see Figs. 4, 5); 5637 ft (1761.6 m) in well 13-5-51-11 (see Fig. 13). Bioturbated remnants of gravel stringer can be seen just above the middle of the core. The bulk of the core shows facies 4, as in Figure 8D. Scale in centimeters.

Figure 12. [A] Gritty siderite at 5667 ft (1770.9 m) in well 13-5-51-11 (see Fig. 13). Note the scattered white grains within and just below the siderite; they are coarser than the grains in the surrounding facies 5 (see Figs. 9A, 13). Internal texture of the siderite suggests some bioturbation before sideritization. Scale in centimeters. [B] Sharp contact of conglomerate on underlying facies 5 at 5641.5 ft (1763.0 m) in well 13-5-51-11 (see Fig. 13). Note gritty layer (center of photo, 2 cm above scale) burrowed into the underlying mudstones. Scale in centimeters.

The conglomerates of the Carrot Creek Member (Plint et al., 1986) rest unconformably and have varying depths of scour on different parts of the "b" and "a" sequences of the Raven River Member (Figs. 4, 5). The conglomerate contact may be sharp (Fig. 12B) or bioturbated (Fig. 12B, C).

Overlying the conglomerates, facies 3P (Fig. 11A), 4P (Fig. 11B), and 5P are locally developed. Following deposition of all the conglomerates and pebble stringers, facies 2 occurs as a "laminated blanket" spread over the entire area. Facies 2, in turn, grades upward into the massive dark mudstones of facies 1, separated from facies 2 by a gritty layer.

These facies sequences are illustrated by the cores shown in Figures 13, 14, and 15.

Cross Sections

The two cross sections in Figures 4 and 5 (located in Figs. 1, 6, 7) illustrate

1. The presence of a major erosion surface, E5

2. The two-dimensional geometry of the conglomerate bodies

3. Vertical and lateral facies relationships

The sections (both log and core) are hung on the stratigraphic marker E7/T7, a pair of induction-log peaks approximately 32 m above the top of the Carrot Creek Member. The lower markers (E4/T4, E3/T3, and E1/T1) are subparallel to the upper markers, and the E1/T1 to E7/T7 separation varies from 92 to 107 m within the field.

Figure 12. [C] Contact of conglomerate on underlying facies 7 at 5430 ft (1696.9 m) in well 2-21-51-11 (see Fig. 13). Note particularly the bioturbation (horizontal side-filled burrow) at the contact. Round object 1 cm above scale is a burrow, not a pebble. Scale in centimeters. [D] Contact of conglomerate on underlying facies 4 at 5386 ft (1683.1 m) in well 10-21-51-11 (see Fig. 15). Note how pebbles have been displaced downward from the base of the conglomerate by bioturbation. Scale in centimeters.

Figure 13. The core of well 13-5-51-11 is representative of sequences from the "terrace" or "bumps." The next four pages of core photographs show the middle 60 ft (18.7 m) of the core; the full section is shown in Figure 5. Core depths are all given in feet; the bottom of the core is in the lower left and the top in the upper right. Coarsening-upward sequence "b" begins below 5689 ft (1777.8 m) and continues to the gritty siderite at 5667 ft (1770.9 m). Sequence "a" begins at 5666.5 ft (1770.8 m) with facies 5; in other cores, the facies above the gritty siderite is finer (facies 3 or 4) and continues into about 16.5 ft (5 m) of hummocky cross-stratified sandstones of facies 7 (which begin at about 5658 ft (1768.1 m). The sandstones are abruptly overlain by conglomerate at 5641.5 ft (1763.0 m). The conglomerates consist mostly of pebbles scattered in bioturbated mudstones (facies 4P, 5641.5 to about 5634 ft (1763.0 to 1760.6 m) and are overlain by the "laminated blanket" of facies 2.

5674

FACIES 5

5659

5667

GRITTY
SIDERITE

13-5-51-11

IMP CYNTHIA SOUTH

5674

FACIES 5

248

FACIES 7

13-5-51-11
IMP CYNTHIA SOUTH
5659

5644

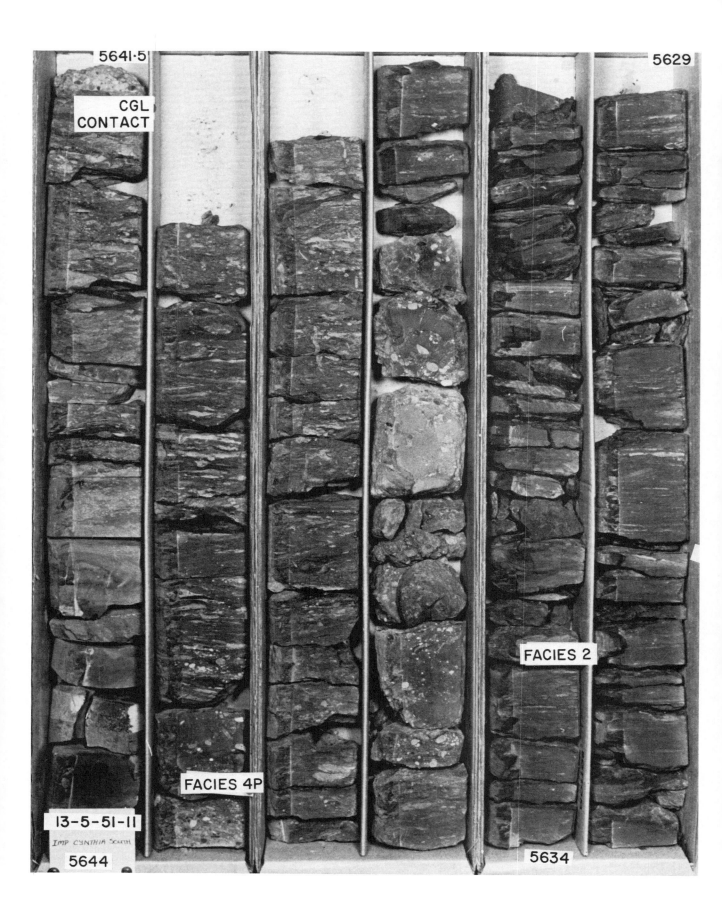

5641·5

CGL
CONTACT

5629

FACIES 2

FACIES 4P

13-5-51-11

IMP CYNTHIA SOUTH

5644

5634

Figure 14.   The core of well 2-21-51-11 is typical of the "bevel," or
the paleo-landward side of a hollow. It is shown in the
next four pages of core photographs; the bottom of the
core is in the lower left, the top in the upper right.
Empty parts of core boxes are due to missing core. The
core begins with facies 5, overlain by a thin veneer of
facies 7 (5432.5 to 5430 ft; 1697.7 to 1696.9 m) and is
abruptly truncated by the conglomerate at 5430 ft (1696.9
m) (Fig. 5). The conglomerate is thick (5430 to 5396 ft;
1696.9 to 1686.3 m), a total of 34 ft or 10.4 m, and
contains a few sandier beds toward the base. It is very
massive and coarser grained toward the top. It is
abruptly overlain by the "laminated blanket" of facies 2,
and facies 3P and 4P are not developed.

TOP

CGL
CONTACT
5430

5425

2-21-51-11

Champlin et al
CYN-PEM

5440

FACIES 5

FACIES 7

FACIES 8

253

2-21-51-11

CHAMPLIN ET AL
CYN-PEN

5425

5395

2-21-51-11
CHAMPLIN ET AL
CSN-PEM
5410

255

5385

FACIES 2

2-21-51-11

CHAMPLIN ET AL
CYN-PEM

5395

Figure 15. The core of well 10-21-51-11 is typical of the seaward sides of hollows. It is shown in the next four pages of core photographs; the bottom of the core is in the lower left, the top in the upper right. The core begins with facies 3 and 4 (5400 to 5386 ft; 1687.5 to 1683.1 m) of sequence "b." The log and core cross sections (Fig. 5) show that the gritty siderite and the "a" sequence are both cut out by the conglomerate at 5386 ft (1683.1 m). However, the conglomerate is poorly developed and is mostly represented by gravel stringers within bioturbated mudstones (5386 to about 5348 ft; 1683.1 to 1671.3 m). Most of this bioturbated section is interpreted to have been deposited during the early stages of transgression, and the gravel stringers represent storm-emplaced layers derived from thicker conglomerates in the "hollows" (e.g., well 2-21-51-11). The "laminated blanket" of facies 2 overlies facies 3P at about 5348 ft (1671.3 m).

5385

CGL
CONTACT

5392
FACIES 4

FACIES 3

10-21-51-11

5400

FACIES 5P

10-21-51-11

5385

5370

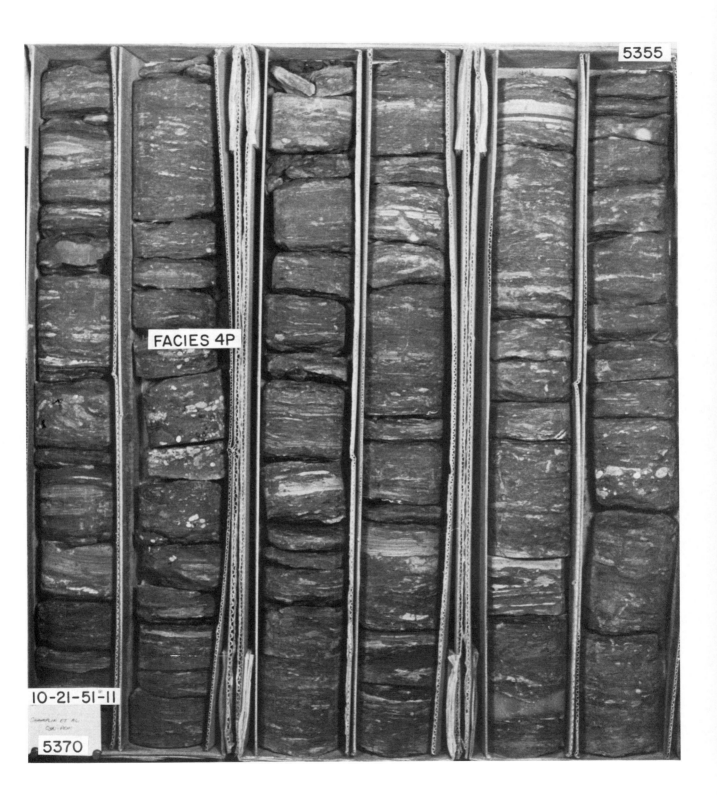

FACIES 4P

10-21-51-11

5370

5355

261

FACIES 2

5348

FACIES 3P

5340

10-21-51-11

5355

Cross Section A

In cross section A (Fig. 4), the base of the conglomerate is clearly erosive, cutting out the hummocky cross-stratified sandstones and part of the "a" sequence between wells 7-34 and 10-35 (about 1.7 km). It continues to cut downward, removing all of the "a" sequence, the gritty siderite, and part of the "b" sequence between wells 10-35 and 6-1 (about 3 km). From the cross section, it is possible to define a *terrace*, where the conglomerate rests on hummocky cross-stratified sandstones. Basinward, the terrace is cut by a *bevel*, and beyond the toe of the bevel is a *hollow*.

The top of the conglomerate is essentially flat across the terrace and the bevel; the conglomerate thins dramatically in the hollow, with only a thin veneer of pebbles in well 6-1. The hollow is filled with bioturbated mudstones containing pebble stringers (facies 3P and 4P). The "laminated blanket" (facies 2) covers the entire sequence.

Cross Section B

Cross section B (Fig. 5) shows topographic relief in the form of *bumps and hollows* basinward of the bevel. Typical cores from the bumps and hollows are shown in Figures 13, 14, and 15. Within the two bumps (wells 13-5/4-16 and 2-28), the preserved facies sequence (Fig. 13) is identical to that underlying the terrace in wells 13-21 and 7-34, namely, sequences "b" and "a" separated by the gritty siderite (Fig. 5). This clearly demonstrates that the bumps represent a remnant erosional topography rather than "ridges" or "bars" that grew after formation of the erosion surface E5. In the hollows, thick conglomerates are banked up against the landward side (Fig. 14) and overlain on the seaward side by Facies 5P, 4P, and 3P (Fig. 15). After emplacement of the pebbles, facies 2 (laminated mudstone) blanketed the entire area.

Relationship of Conglomerates to the Erosion Surface

The cross sections (Figs. 4, 5) illustrate the presence of a major erosion surface (E5). The shape of this surface has been determined by isopaching the interval between the datum (E7/T7) and the base of the conglomerate (Figs. 4, 5), using both core and well-log data. The resulting map is shown in Figure 6; the contour values represent depths below the marker--the higher the isopach value, the greater the depth of scour.

The terrace is a broad, irregularly undulating expanse that lies in the southwestern part of the area. The bevel is a narrow belt where the hummocky cross-stratified sandstones of facies 7 are truncated. Basinward (northeast) of the bevel, there is a topography of bumps and hollows; these are long and narrow, and trend slightly oblique to the bevel. The maximum erosional relief is about 20 m.

Core and log data have been used to construct the conglomerate isopach map (Fig. 7), which shows a series of pods with thicknesses up to 19 m. The pods are oriented more or less parallel to the hollows. Between the pods is a continuous veneer of conglomerate generally less than a meter thick.

Superimposing Figure 7 on Figure 6 shows that the thicker conglomerates generally coincide with the deeper scours on the erosion surface. Not all deep scours contain conglomerate, as shown by well 10-21 (Figs. 5, 15), where the hollow is seaward of a thick conglomerate, and the fill is mudstones with pebble stringers.

Conclusions

1. Sequence "b" represents a progressive shallowing-upward sequence through facies 1, 3, 4, and 5.

2. The gritty siderite separating sequences "b" and "a" results from a pause in deposition, presumably due either to a minor drop in sea level or a stillstand.

3. Sequence "a" represents another progressively shallowing-upward sequence through bioturbated mudstones of facies 4 and 5, culminating in hummocky cross-stratified sandstones, suggesting deposition above storm wave base.

4. A major drop in relative sea level resulted in the erosion surface, E5. This surface is believed to have been formed by wave scouring, which reestablished a new shoreface (bevel) basinward of the terrace. The bumps and hollows seaward of the bevel represent a remnant erosional topography. The orientation and geometry of the surface (particularly the remnant topography of bumps and hollows, and the extensive bevel) preclude an origin by subaerial fluvial erosion during lowstand.

5. This sea level drop also caused lowering of the base level and hence the incision of rivers, allowing the transport of coarse material to the newly established shoreface (bevel).

6. During a stillstand period, thick clast-supported conglomerate (up to 19 m) was deposited against the shoreface (bevel) and on the landward side of the hollows.

7. The seaward sides of the hollows were filled with mudstones deposited during the transgression. During storms, gravel was reworked to form scattered pebbles and pebble stringers (facies 3P,

4P, and 5P) in the mudstones seaward of the thick gravels. The transgression also reworked gravel back across the terrace from the thick accumulations against the bevel.

8.   With continued transgression, a laterally continuous and widespread "laminated blanket" (facies 2) was deposited over all underlying conglomerates and pebbly mudstones.

9.   After deposition of this laminated blanket, widespread deposition of massive mudstones (facies 1) occurred.

Acknowledgments

We thank the Home Oil Company for the logistical and technical support of this study, in particular George Fong and Sid Leggitt. Financial support was provided by the Natural Sciences and Engineering Research Council of Canada in the form of Strategic and Operating grants to R. G. Walker and a graduate scholarship to K. M. Bergman. We particularly thank Art Shepard of the Energy Resources Conservation Board of Alberta for allowing use of the cores in this workshop. Jack Whorwood prepared the photographs. We would like to thank Guy Plint for ideas and comments on the manuscript.

References

Caldwell, W. G. E., 1984, Early Cretaceous transgressions and regressions in the Southern Interior Plains, in D. F. Stott and D. J. Glass (eds.), The Mesozoic of Middle North America: Can. Soc. Petrol. Geol., Mem. 9, p. 173-203.

Duke, W. L., 1985, Sedimentology of the Upper Cretaceous Cardium Formation in southern Alberta, Canada: Ph.D. thesis, McMaster University, Hamilton, Canada, 724 p.

Jeletzky, J. A., 1978, Causes of Cretaceous oscillations of sea level in western and arctic Canada and some general geotectonic implications: Geol. Surv. Canada, Paper 77-18, 38 p.

Kauffman, E. G., 1969, Cretaceous marine cycles of the Western Interior: Mountain Geol., v. 6, p. 227-45.

Kauffman, E. G., 1977, Geological and biological overview, Western Interior Cretaceous Basin: Mountain Geol., v. 14, p. 129-52.

Keith, D. A. W., 1985, Sedimentology of the Cardium Formation, Willesden Green Field, Alberta: M.S. thesis, McMaster University, Hamilton, Canada, 233 p.

Krause, F. F., and Nelson, D. A., 1984, Storm event sedimentation: lithofacies association in the Cardium Formation, Pembina area, west-central Alberta, in D. F. Stott and D. J. Glass (eds.), The Mesozoic of Middle North America: Can. Soc. Petrol. Geol., Mem. 9, p. 485-511.

Plint, A. G., Walker, R. G., and Bergman, K. M., 1986, Cardium Formation 6. Stratigraphic framework of the Cardium in subsurface: Bull. Can. Petrol. Geol., v. 34 (in press).

Raddysh, H., 1986, Sedimentology of the Viking Formation at Gilby A and B Fields, Alberta: B.S. thesis, McMaster University, Hamilton, Canada.

Swagor, N. S., Oliver, T. A., and Johnson, B. A., 1976, Carrot Creek field, central Alberta, in M. M. Lerand (ed.), The sedimentology of selected clastic oil and gas reservoirs in Alberta: Can. Soc. Petrol. Geol., p. 78-95.

Swift, D. J. P., and Niedoroda, A. W., 1985, Fluid and sediment dynamics on Continental Shelves, in R. W. Tillman, D. J. P. Swift, and R. G. Walker (eds.), Shelf sands and sandstone reservoirs: Soc. Econ. Pal. Min., Short Course 13, p. 47-133.

Walker, R. G., 1983a, Cardium Formation 1. "Cardium a turbidity current deposit" (Beach, 1955): a brief history of ideas: Bull. Can. Petrol. Geol., v. 31, p. 205-12.

Walker, R. G., 1983b, Cardium Formation 3. Sedimentology and stratigraphy in the Garrington-Caroline area, Alberta: Bull. Can. Petrol. Geol., v. 31, p. 213-30.

Weimer, R. J., 1984, Relation of unconformities, tectonics and sea-level changes, Cretaceous of Western Interior U.S.A., in J. S. Schlee (ed.), Interregional unconformities and hydrocarbon accumulation: Amer. Assoc. Petrol. Geol., Mem. 36, p. 7-35.

Wright, M. E., and Walker, R. G., 1981, Cardium Formation (Upper Cretaceous) at Seebe, Alberta--storm-transported sandstones and conglomerates in shallow marine depositional environments below fairweather wave base: Can. Jour. Earth Sci., v. 18, p. 795-809.

MARINE-SHELF BAR SAND/CHANNELIZED SAND SHINGLED COUPLET,
TERRY SANDSTONE MEMBER OF PIERRE SHALE,
DENVER BASIN, COLORADO

Charles T. Siemers and John H. Ristow

Sedimentology, Inc., Boulder, Colorado 80302

Abstract

The Terry Sandstone Member, which lies within the middle part of the Upper Cretaceous Pierre Shale in the Denver basin of northeastern Colorado, has been interpreted as a marine-shelf depositional sequence. Such interpretation has been based mainly on the sedimentological character of the sandstone units within the depositional sequence and on their position seaward of correlative Lower Cretaceous shorelines. This study conducted a process sedimentological analysis of the Terry Sandstone sequence in an area the size of two townships approximately 8 mi downdip from the sandstone outcrop. Four core sequences, two of which are illustrated herein, and wire-line logs from over 200 wells were utilized to construct a series of subsurface maps and cross sections with which to evaluate sandstone body geometry, trend, and sedimentological character.

The Terry Sandstone Member thickens over a distance of 11 mi from less than 60 ft thick in the northwestern corner of the study area to greater than 120 ft thick in the southeastern corner of the study area. An upper, sandy mudstone sequence forms a blanketlike unit that ranges from 50 to 70 ft thick throughout the study area. The lower sandstone unit of the Terry Sandstone is composed of two distinct northwest-trending, elongate, lens-shaped sandstone bodies (designated herein as the "northeastern" and "southwestern" sandstone bodies) that display a shinglelike overlapping relationship across a 1- to 2-mi wide, northwest-trending strip. The two units are separated by a shale permeability barrier unit (designated herein as the shale "notch").

Within the lower unit of the Terry Sandstone Member, the northeastern sandstone body displays a symmetrical lens-shaped cross section up to 70 ft thick and 4-5 mi in maximum width. It is characterized by a gradational base, a coarsening-upward vertical sequence, and a sharp upper contact with overlying marine mudstone. The shale "notch" appears to be composed of silt-poor clay shale; it "migrates" from the top to the base of the lower sandstone unit of the Terry Sandstone within 1.5 mi and effectively separates the two sandstone bodies. The Southwestern Sandstone Body displays an asymmetrical lens-shaped cross section (with more rapid thinning along the northeastern flank), is up to 50 ft thick and 5-6 mi in maximum width, and is characterized by a sharp, erosional basal contact, fining-upward vertical sequence, and gradational top.

Evidence suggests that both sandstone bodies of the lower sandstone unit of the Terry Sandstone were deposited in a shallow marine-shelf setting. This evidence includes the general paleogeographic setting and features such as the extensively burrowed to bioturbated sandstone and mudstone sequences, the abundance of glauconite, and the extensive marine mudstone intervals of the Pierre Shale that completely encase the Terry Sandstone sequence. The northeastern sandstone body formed as an offshore marine sand bar, whereas the southwestern sandstone body apparently represents a channelized deposit composed of active channel-fill sand and overlying inactive channel-fill sandy mudstone. Of economic significance is the segregation of hydrocarbon types in the sandstone bodies; noncommercial volumes of gas are present in the northeastern sandstone body and commercial volumes of oil in the southwestern sandstone body.

Introduction

The Upper Cretaceous Terry Sandstone Member of the Pierre Shale, a hydrocarbon-productive sandstone reservoir unit within the Denver basin of Colorado (Figs. 1, 2), has been considered by several previous investigators to represent nearshore (paralic) to shelf marine siliciclastic deposition (e.g., Moredock and Williams, 1976; Kiteley, 1977, 1978; Porter and Weimer, 1983). Most investigators, including exploration and production geologists, have generally applied the rather conventional offshore bar model as an explanation of origin of the Terry Sandstone bodies and as a tool in subsurface exploration for those sandstone bodies. Although significant amounts of oil and gas have been discovered and developed using such conventional models, this paper suggests that the present depositional and exploration model for the Terry Sandstone needs some modification.

In 1982, the senior author's initial examination of a hydrocarbon-productive interval of the Terry Sandstone sequence of the Denver basin revealed a sandstone body is strikingly anomalous in terms of application of the coarsening-upward offshore bar model. Examination of the Brooks Expl. Inc. Firestien No. 1-30 core sequence (presented herein) and two similar core and associated wire-line log sequences

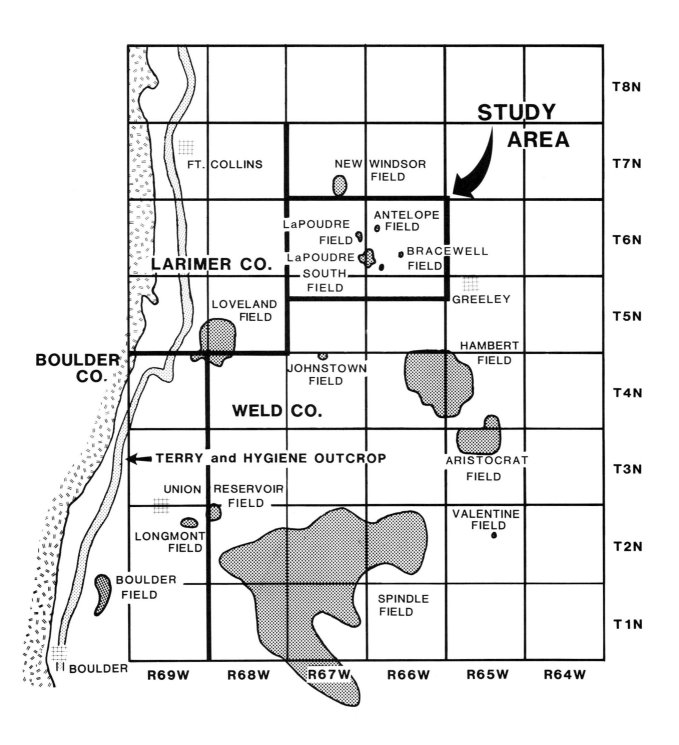

Figure 1. Locality map showing fields presently producing from the Terry and Hygiene Sandstone members of the Pierre Shale in the northern part of the Denver basin, Colorado.

AGE	FORMATIONS			THICKNESS (Ft)	
TERTIARY	Denver-Arapahoe Fm.			500	
	Laramie Fm.				
	Fox Hills Fm.			60	
UPPER CRETACEOUS	Pierre Sh	Terry (Sussex) Sandstone		8000	●
		Hygiene (Shannon) Sandstone			●
		Mitten Member			
		Sharon Springs Member			●
	Niobrara	Smoky Hill Member		350	
		Fort Hays Limestone			●
		Codell Sandstone		420	●
		Carlile Shale			●
		Greenhorn Limestone			●
		Graneros Shale			●
LOWER CRETACEOUS	Dakota Grp	"D" Sand	Huntsman/Mowry	370	●
		"J" Sand			●
		Skull Creek Shale			
		Plainview Sandstone			
		Lytle Formation (Lakota Sandstone)			●
JURASSIC	Morrison Formation			475	
	Ralston Creek Formation				
TRIASSIC	Lykins Formation			400	
PERMIAN	Lyons Formation			120	●
PENN	Fountain Formation			1000	

Figure 2. Generalized column of major stratigraphic units of the Front Range area of Colorado. Dots indicate the numerous oil and gas productive units of the Denver basin (modified from Weimer, 1973).

272

through the sandstone unit revealed a sharp basal contact, laminated to ripple-bedded, and highly burrowed to bioturbated sandstone in the lower part grading upward to sandy mudstone and overlain gradationally by marine mudstone of the Pierre Shale. Such a sequence obviously is in strong contrast to the conventional offshore marine bar model, which displays a gradational base, coarsening-upward textural trend, and sharp upper contact with overlying marine mudstone.

For this paper, and the SEPM core workshop, investigation of the Terry Sandstone sequence was expanded to include detailed examination of more than 200 wire-line log sequences of the Terry Sandstone throughout an area the size of two townships (72 mi^2) in Weld County, Colorado (Figs. 1, 3). Also included is the sedimentological analysis of an additional significant core sequence (i.e., that from the Brooks Expl. Inc., Leffler No. 1 well). This investigation has provided not only a better understanding of the anomalous fining-upward sandstone core sequences of the Terry Sandstone, but also an interesting depositional sequence of two distinct marine-shelf sandstone bodies. Significant aspects of these two northwest-trending, elongate sandstone bodies include the following: (1) they formed by significantly different depositional processes; (2) they display a shinglelike lateral and vertical geometric relationship; and (3) they are separated by an impermeable shalelike barrier. One of the sandstone bodies is oil productive, whereas the other is a nonproductive gas-bearing unit, which has added considerable economic significance to this study. A core sequence through the oil-productive sandstone body (Brooks Expl. Inc., Firestien No. 1-30 well) is discussed in this paper and will be available for examination during the core workshop. An additional core sequence

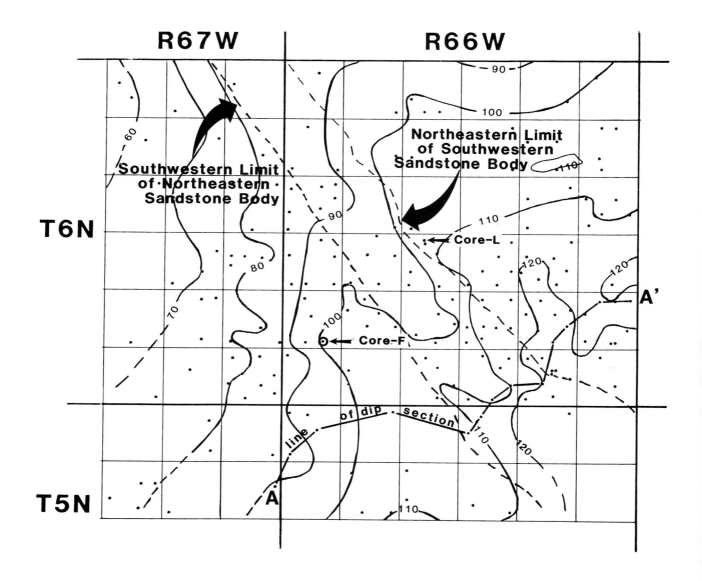

Figure 3. Isopach map of the Terry Sandstone Member of the Pierre Shale. Note location of cored wells (core F - Brooks Expl. Inc., Firestien No. 1-30; core L - Brooks Expl. Inc., Leffler No. 1) and position of wire-line log section A-A' approximately parallel to depositional dip (see Fig. 8). Also note approximate limits of the two separate Terry Sandstone bodies present in the northeastern and southeastern parts of the study area and separated by a shale-like permeability barrier. The area between the northwest-trending depositional limits is an area of overlap of the two sandstone bodies, as well as the area where the shalelike permeability barrier is best developed.

through the gas-bearing unit (Brooks Expl. Inc., Leffler No. 1 well) was acquired for the core workshop to provide detailed comparison and contrast of the two sandstone bodies of the lower unit of the Terry Sandstone.

Objectives

The primary objective of this study was to describe and discuss the origin of the Terry Sandstone sequence, which formed in an ancient marine-shelf depositional setting. Since one of the sandstone bodies delineated in the study is oil productive, development of a genetic-predictive model for the origin of the depositional sequence and for its performance as a hydrocarbon reservoir also has economic significance. The study is being expanded to cover a broader area of the Denver basin and to investigate in more detail the petrophysical aspects of the sandstone reservoir; therefore, this paper represents only a preliminary report of a larger investigation. A secondary objective of this presentation is to demonstrate the effectiveness of the process sedimentology approach to core and wire-line log investigation of subsurface depositional sequences. That approach can be contrasted with the less effective, but more commonly utilized, model approach.

Stratigraphic Terminology

The Terry and Hygiene Sandstone members of the Pierre Shale in the Denver basin of Colorado have been, and are, commonly referred to as the "Sussex" and "Shannon" members, respectively. The Sussex and Shannon sandstone members are well-known marine-shelf sandstone units of the Pierre Shale in the Powder River basin of eastern Wyoming (see Hobson et al., 1982; Tillman and Martinsen, 1984). It is important to note, however, that the Denver basin Pierre Shale sandstone members are stratigraphically younger than those of the Powder River basin (Kiteley, 1976; Porter, 1976). To avoid confusion, the names "Terry Sandstone" and "Hygiene Sandstone" are throughout this paper.

Hydrocarbon Production

The Terry and Hygiene Sandstone members have been producers of oil and gas in Weld County, Colorado, and continue to be important exploration and development targets in the northern Denver basin. These sandstone reservoirs are generally lenticular and discontinuous, and diagenetic clays frequently complicate drilling, completion, and production programs. Economic production, however, has continued since the discovery of Boulder field in 1901. Except for a flurry of unsuccessful exploration activity in the 1950s and 1960s, significant exploration for and production of Terry and Hygiene hydrocarbon have occurred only since the early 1970s, when Spindle field was established (Moredock and Williams, 1976). Much of the exploration activity since that time, including many of the recent oil discoveries, was carried out using limited data and rather conventional models of sandstone reservoir character, trend, distribution, and origin. That exploration activity generated a large amount of data utilized in this study to develop a more detailed genetic model for the Terry Sandstone Member of the middle Pierre Shale in the northern part of the Denver basin of Colorado.

Methodology

The Terry Sandstone core sequence from the Brooks Expl. Inc., Firestien No. 1-30 well was handled and analyzed sedimentologically according to procedures delineated in Siemers and Tillman (1981). The core was relabeled, slabbed, lapped (to remove saw marks), and photographed prior to description. A slabbed core interval of the Terry Sandstone sequence from the Brooks Expl. Inc., Leffler No. 1 well is presently stored in the U.S. Geological Survey core library in Denver, Colorado, and that core was photographed and described at that facility. Complete sets of photographic prints of the continuous core sequences are included in Figures 6 and 9. A binocular microscope was used to describe both cores lithologically; grain-size determinations were made with the aid of a grain-size comparison strip, and rock color determined by comparison with the GSA rock color chart. Detailed continuous lithologic sketches (Figs. 5, 8) and brief written descriptions are included herein. A core-to-log depth correction was established and the wire-line log sequence calibrated to the lithologic significance of various log characteristics (Figs. 4, 7).

Subsurface maps and cross sections of the Terry Sandstone in the study area were made using standard subsurface mapping procedures. Gamma-ray and resistivity logs were obtained for most of the 203 well localities used in the study. Initial critical procedures included determination of the various log picks and recognition of the three major log pattern types shown in Figure 7. Correlation of the wire-line logs began with the core-calibrated log from the Brooks Expl. Inc., Firestien No. 1-30 well (Fig. 4). Once we determined that two distinct and separate sandstone bodies are present within the lower unit of the Terry

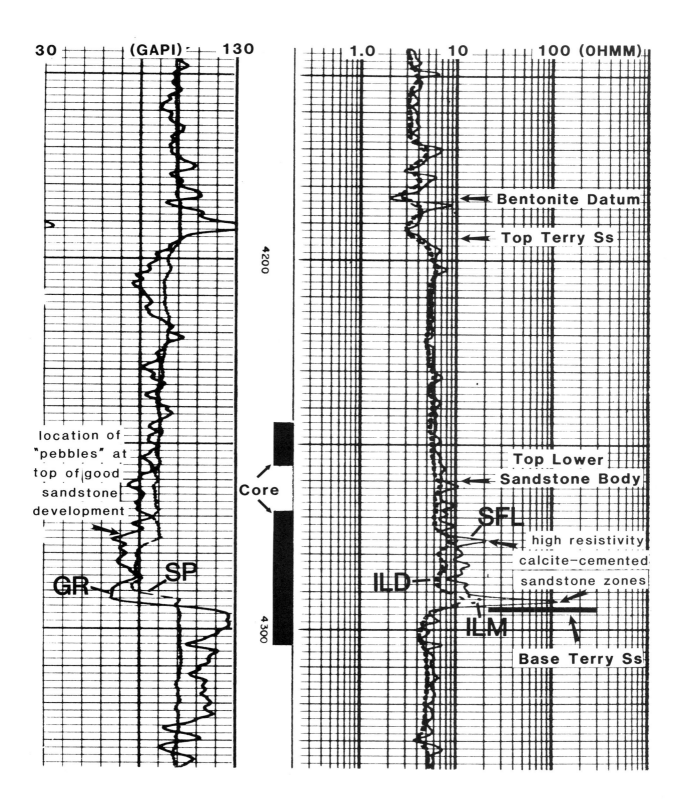

Figure 4. Wire-line logs through the Terry Sandstone sequence in the Brooks Expl. Inc., Firestien No. 1-30 well (SW-SE Sec. 30, T6N R66W, Weld Co., Colorado). Note position of log picks and cored interval (core-to-log depth correction is approximately +4 ft).

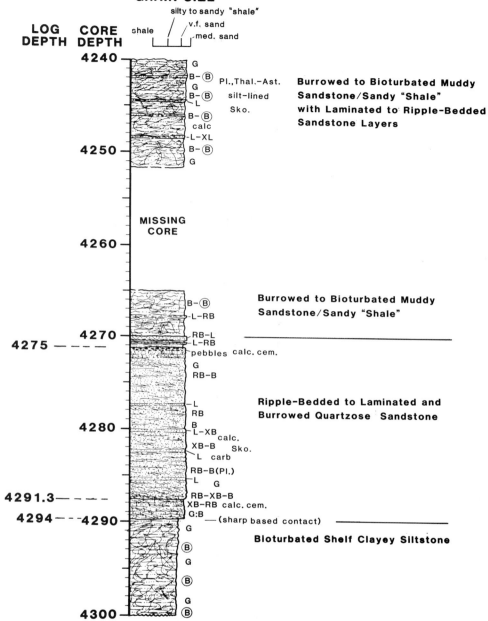

Figure 5. Sketch of lithology for core from the Terry Sandstone sequence of the Brooks Expl. Inc., Firestien No. 1-30 well (SW-SE Sec. 30, T6N R66W, Weld Co., Colorado). Note core-to-log depth correction of +4 ft (and log-to-core depth correction of -4 ft). Symbols along right margin indicate the following: L = laminated; RB = ripple-bedded; XB = cross-bedded; B = burrowed; B = bioturbated; G = glauconite (tr-5%); Pl. = *Planolites* burrows; Thal. = *Thalassinoides* burrows; Ast. = *Asterosoma* burrows; Sko. = *Skolithos* burrows.

279

Figure 6. Photographs of slabbed and lapped whole-diameter core from the Terry Sandstone sequence and upper part of underlying Pierre Shale, Brooks Expl. Inc., Firestien No. 1-30 well (SW-SE Sec. 30, T6N R66W, Weld Co., Colorado). Note core-to-log depth correction of +4 ft (6 pages).

BROOKS #1-30 FIRESTEIN
WELD COUNTY, COLORADO

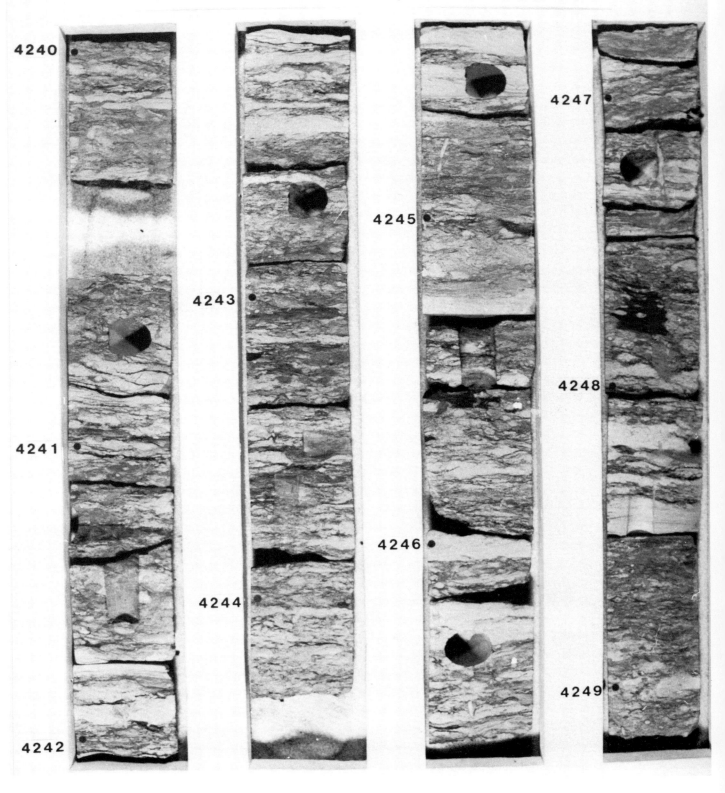

4240

4243

4245

4247

4241

4248

4246

4244

4242

4249

BROOKS #1-30 FIRESTEIN
WELD COUNTY, COLORADO

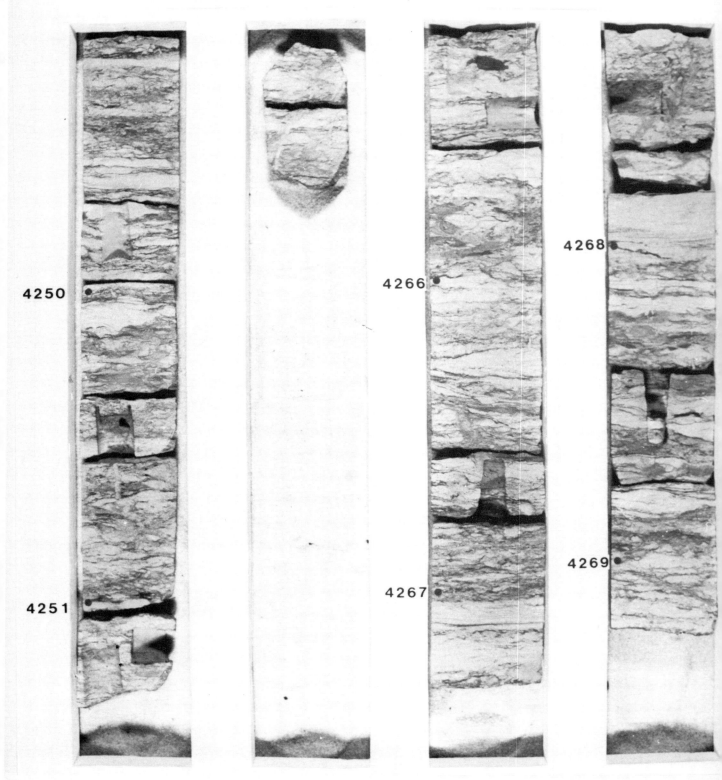

4250

4251

4266

4267

4268

4269

BROOKS #1-30 FIRESTEIN
WELD COUNTY, COLORADO

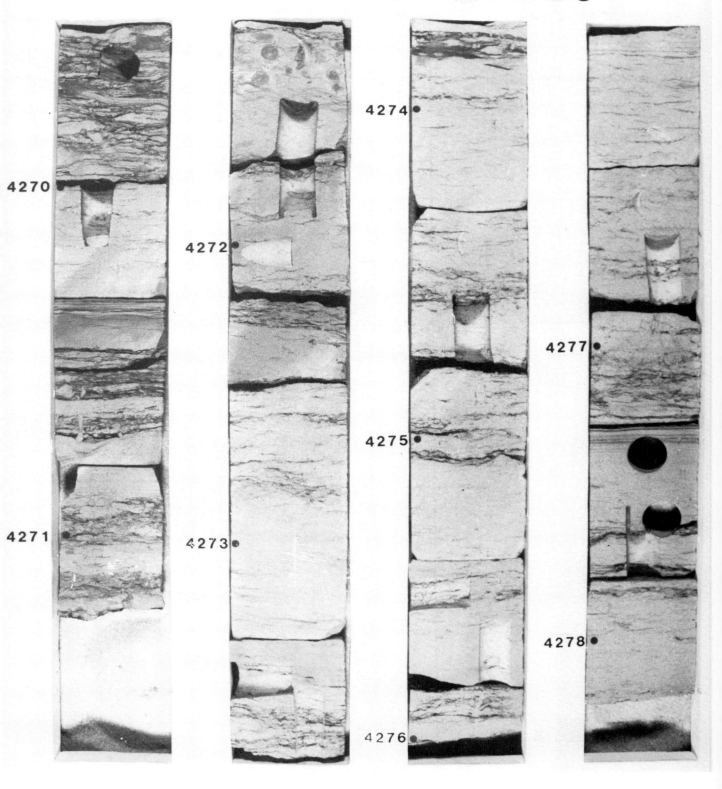

BROOKS #1-30 FIRESTEIN
WELD COUNTY, COLORADO

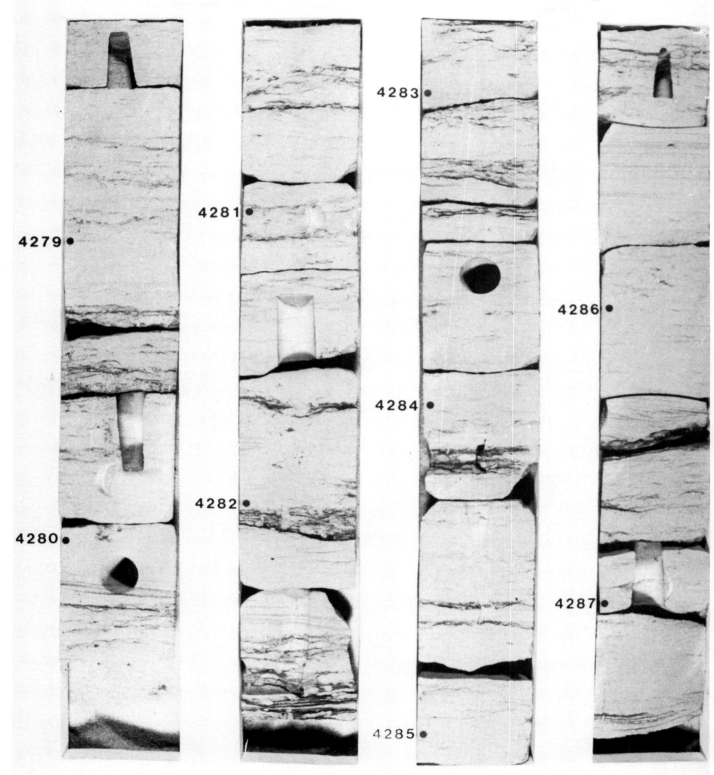

BROOKS #1-30 FIRESTEIN
WELD COUNTY, COLORADO

4288

4289

4290

4291

4292

4293

4294

4295

BROOKS #1-30 FIRESTEIN
WELD COUNTY, COLORADO

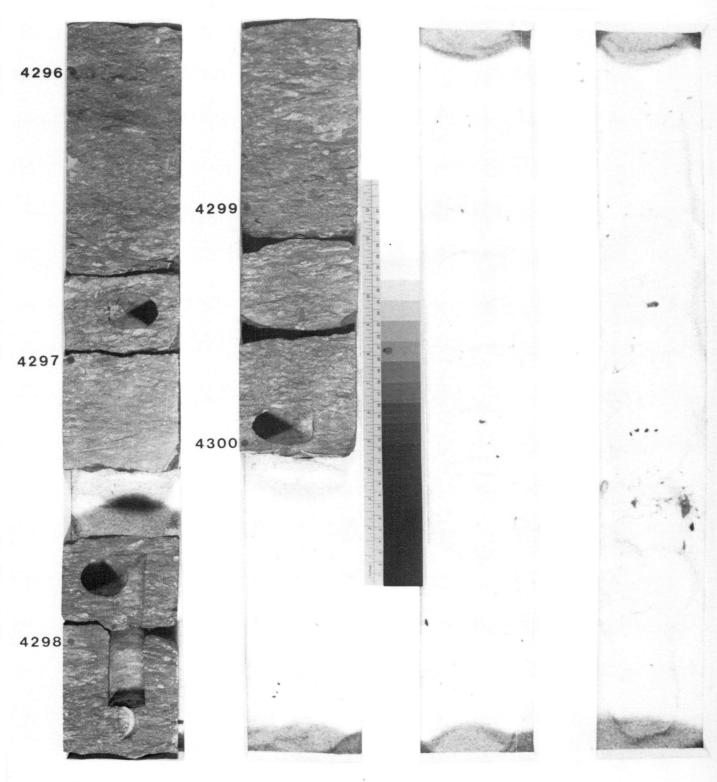

4296

4297

4298

4299

4300

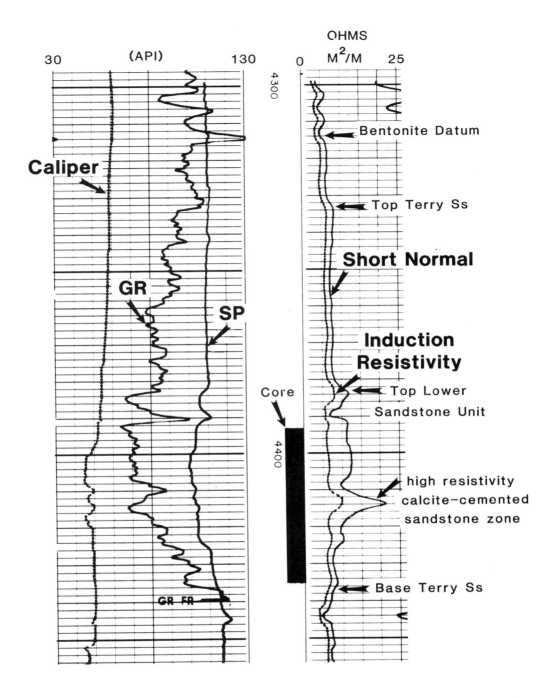

Figure 7. Wire-line logs through the Terry Sandstone sequence in the Brooks Expl. Inc., Leffler No. 1 well (NE-NW Sec. 21, T6N R66W, Weld Co., Colorado). Note position of log picks and the core interval. Bar showing core interval has been corrected to log depth (core-to-log depth correction is approximately +13 ft). Thin shale bed noted at log depth 4388-4390 ft represents the top of the coarsening-upward northeastern sandstone body of the lower part of the Terry Sandstone in the study area. The sandstone interval at approximate log depth 4383-4388 ft represents a thin interval of the southwestern sandstone body along its northeastern margin, where it overlaps the northeastern sandstone body.

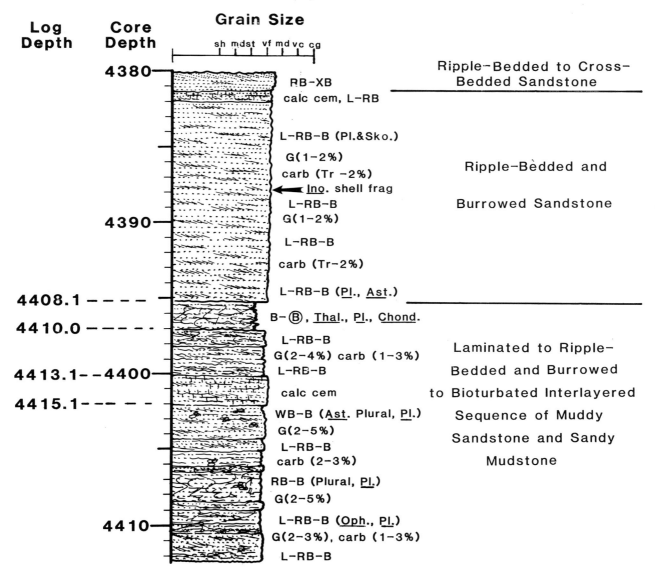

BROOKS EXPL. CO.
Leffler No. 1

NE NW 21-6N-66W
Weld Co., Colorado

Figure 8. Sketch of lithology for core from the Terry Sandstone sequence of the Brooks Expl. Inc., Leffler No. 1 well (NE-NW Sec. 21, T6N R66W, Weld Co., Colorado). Note core-to-log depth correction of +13 ft (and corresponding log-to-core depth correction of -13 ft). Symbols along right margin indicate the following: L = laminated; WB = wavy-bedded; RB = ripple-bedded; XB = cross-bedded; B = burrowed; G = glauconite; carb = carbonaceous debris; Ino. = *Inoceramus*; Pl. = *Planolites*; Thal. = *Thalassinoides*; Ast. = *Asterosoma*; Chond. = *Chondrites*; Plural = plural-curving-tubes; Sko. = *Skolithos*; Oph. = *Ophiomorpha*.

Figure 9. Photographs of slabbed whole-diameter core from the Terry Sandstone sequence, Brooks Expl. Inc., Leffler No. 1 well (NE-NW Sec. 21, T6N R66W, Weld Co., Colorado). Note core-to-log depth correction of +13 ft (3 pages).

BROOKS EXPL.
NO. 1 LEFFLER

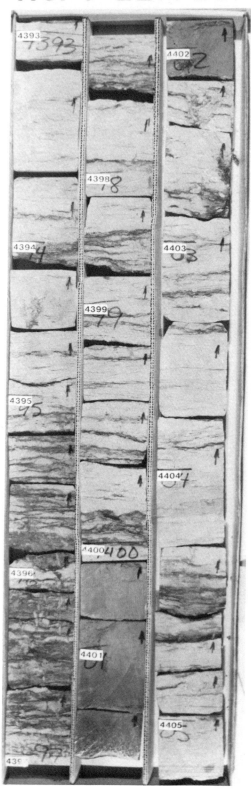

BROOKS EXPL.
NO. 1 LEFFLER

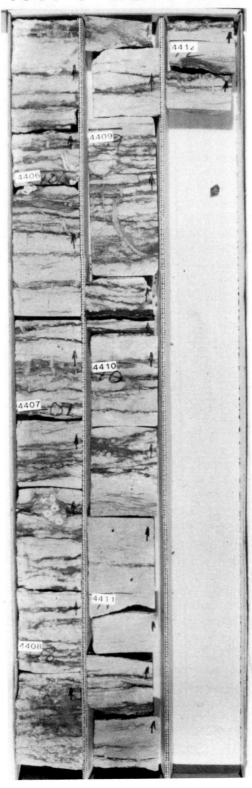

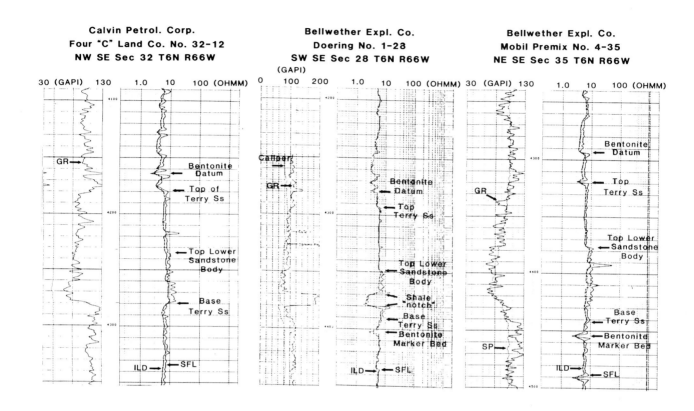

Figure 10. Characteristic ("end member") gamma-ray/resistivity wire-line log patterns for the Terry Sandstone sequence within the study area. Log pattern at left is typical of the sandstone body in the southwestern part of the study area, whereas that at right is typical of the sandstone body in the northeastern part of the study area. Log pattern at center is representative of the central, northwest-trending area where the two separate sandstone bodies overlap. The shale notch (10-11 ft thick in this well) separates the two sandstone bodies vertically and laterally and apparently acts as an effective permeability barrier. The shale notch thins to a zero edge within less than 1 mi to the northeast and southwest of the location of the well shown above (see Fig. 14). The Terry Sandstone sequence below the shale notch represents the southwestern extent of the northeastern sandstone body (e.g., log at right), whereas the Terry Sandstone sequence above the shale notch represents the northeastern extent of the southwestern sandstone body (e.g., log at left).

Sandstone of the study area, the specific set of subsurface maps and cross sections included in this paper (Figs. 11-16) were selected to demonstrate the critical geometric aspects of the sandstone bodies and the shalelike permeability barrier that separates them. Finally, the core sequence from the Brooks Expl. Inc., Leffler No. 1 well was acquired for examination and inclusion in the core workshop.

Process Sedimentology

The process sedimentology approach to the interpretation of the origin of sedimentary sequences emphasizes the recognition and inter-pretation of genetic units and their association in vertical and lateral sequences of sedimentary deposits. It attempts to delineate the phys-ical, biological, and chemical processes that must have occurred to gen-erate the observed sedimentary product (Siemers and Tillman, 1981). Genetic units are lithologic units that display features originating from processes that were relatively constant, or that varied in a uniform manner, throughout the formation of the unit. Major types of genetic units include sedimentation units, ichnogenetic (i.e., trace fossil) units, soft-sediment-deformation (ssd) units, and diagenetic units (see Siemers and Tillman, 1981, for an expanded explanation of these units).

Application of the process sedimentology approach in the analysis of the Terry Sandstone sequence was critical in terms of (1) initial rec-ognition of the significance of the anomalous fining-upward vertical tex-tural trend of the sandstone sequence in the Brooks Expl. Inc., Firestien No. 1-30 core, (2) subsequent recognition and understanding of the significance of the three major log-pattern types present within the Terry Sandstone of the study area, and (3) final determination of

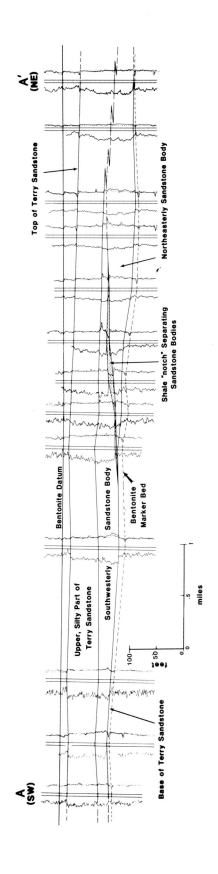

Figure 11. Wire-line log (gamma-ray/resistivity) cross section A-A' of Terry Sandstone sequence oriented approximately normal to sandstone body trends (i.e., approximately parallel to depositional dip). Log locations have been projected into a straight line of section to provide a more exact illustration of sandstone body geometry. Note the manner in which the shale notch separates vertically and laterally the sandstone bodies of the northeastern and southwestern parts of the study area. Line of section is shown in Figures 3 and 14.

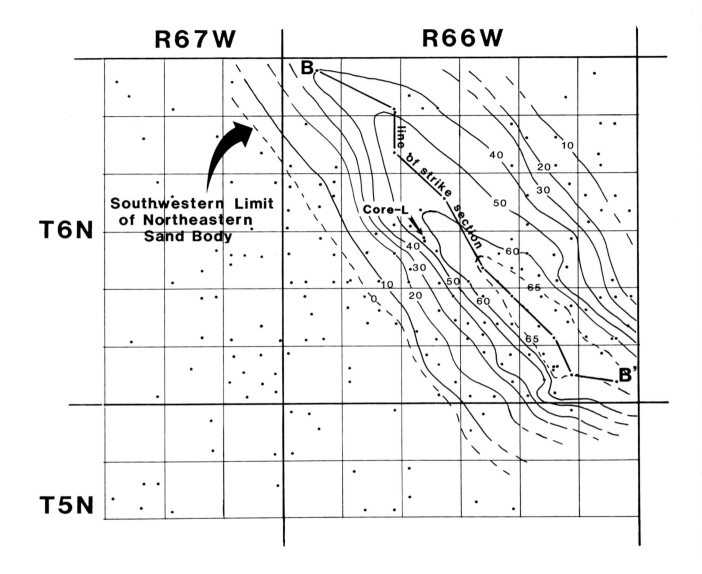

Figure 12. Isopach map of the lower sandstone development of the Terry Sandstone sequence in the northeastern part of the study area. This sandstone body occurs below and northeast of the shale notch unit that separates vertically and laterally the two sandstone bodies of the Terry Sandstone in the study area. Note the generally good cross-sectional symmetry of this northwest-trending sandstone body. This sandstone body also is characterized by a gradational base, coarsening-upward textural trend, and sharp top, which probably represents an offshore marine sandbar development. Note the line of strike wire-line log section B-B' shown in Figure 13, and the location of the Brooks Expl. Inc., Leffler No. 1 core (core-L).

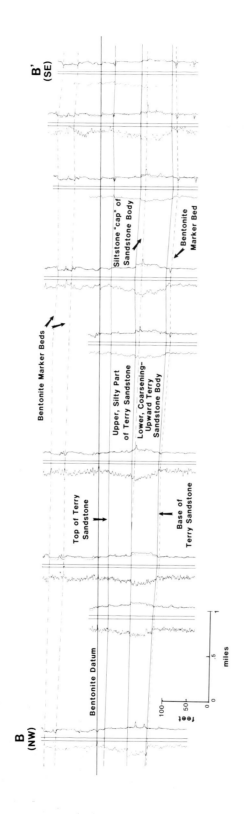

Figure 13. Strike (northwest-southeast) wire-line log (gamma-ray/resistivity) cross section of the northeastern sandstone body (see Fig. 12). Log locations have been projected into a straight line of section to provide a more accurate representation of sandstone body geometry. Note the following: (a) general southeasterly thickening of the total Terry Sandstone sequence owing mainly to thickening of the northeastern sandstone body, (b) gradational base and distinct coarsening-upward textural trend of the northeastern sandstone body, (c) subtle siltstone "cap" at top of the sandstone body as it thickens toward the southeast, and (d) slight increase, in a southeasterly direction, of the interval between base of the sandstone body and bentonite marker bed below.

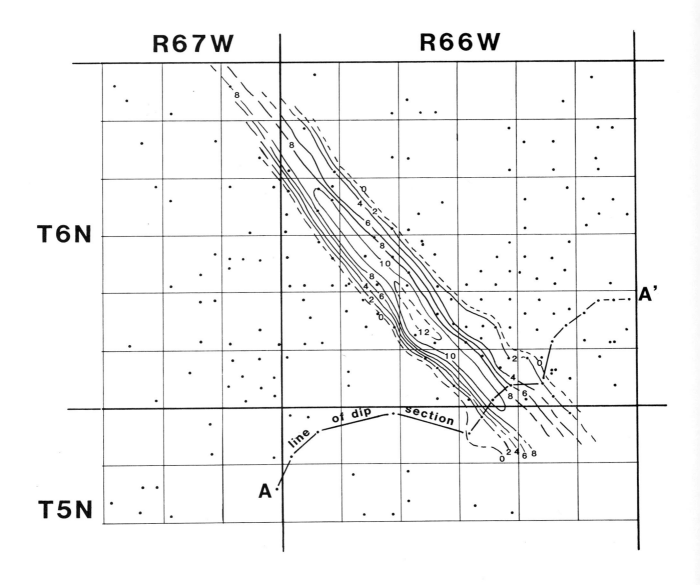

Figure 14. Isopach map of the shale notch unit that vertically and laterally separates the two sandstone bodies of the lower unit of the Terry Sandstone. Note the slight asymmetry of the shalelike unit with more rapid thinning toward the southwest. Such thinning may be due in part to partial erosion prior to deposition of the overlying and south-westerly adjacent sandstone body. Note also the location of wire-line log dip section A-A' shown in Figure 11.

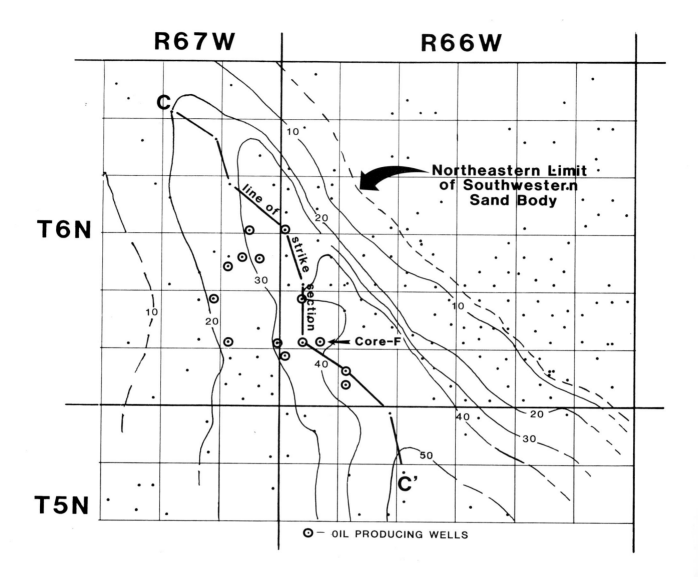

R67W   R66W

T6N

T5N

Northeastern Limit of Southwestern Sand Body

line of strike section

Core-F

C

C'

⊙ — OIL PRODUCING WELLS

Figure 15. Isopach map of the southwestern sandstone body of the lower unit of the Terry Sandstone member in the study area. This sandstone body occurs above and southwest of the shale notch unit that vertically and laterally separates the two sandstone bodies of the lower unit of the Terry Sandstone in the study area. Note the distinct cross-sectional asymmetry of this northwest-trending sandstone body with rapid thinning toward the northeast. This sandstone body also is characterized by a sharp base, fining-upward textural trend, and gradational top, which probably represents a channel-like sandbody development slightly landward of the northeastern sandstone body. The location of this channel-like sandbody was apparently strongly influenced by the preexisting offshore bar-type sandbody to the northeast. Note the location of the line of strike wire-line log section C-C' shown in Figure 16 and the location of the Brooks Expl. Inc., Firestien No. 1-30 core (core-F).

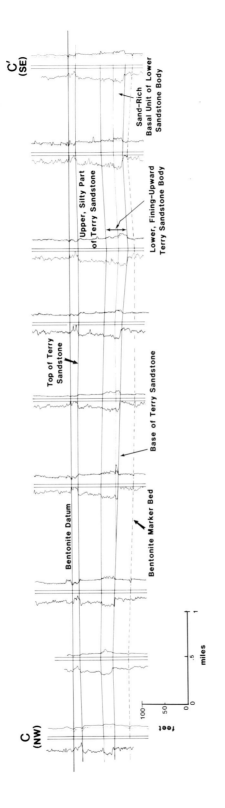

Figure 16. Strike (northwest-southeast) wire-line log (gamma-ray/resistivity) cross section of the southwestern sandstone body of the lower unit of the Terry Sandstone sequence in the study area (see Fig. 15). Log locations have been projected into a straight line of section to provide a more exact representation of sandstone body geometry. Note the following: (a) general southeasterly thickening of the total Terry Sandstone sequence owing mainly to thickening of the southwestern sandstone body, (b) sharp base and distinct fining-upward textural trend of the southwestern sandstone body, (c) southeastward-thickening, sand-rich basal unit of the southwestern sandstone body, and (d) distinct decrease, in a southeasterly direction, of the interval between the base of the sandstone body and the bentonite marker bed below.

301

probable origin of the overall depositional sequence of the Terry Sandstone within the study area.

It is significant that the core sequence used in this study from the Brooks Expl. Inc., Leffler No. 1 well was obtained after overall analysis of the Terry Sandstone was completed within the study area. The lithologic sequence of the lower unit of the Terry Sandstone within the Leffler No. 1 core is precisely that predicted from the analysis of all other subsurface data, and it confirms the validity of the process sedimentology approach to genetic analysis of subsurface sedimentary sequences.

Terry Sandstone Core Sequences

Initial critical procedures in the geological analysis of the Terry Sandstone depositional sequence involved (1) lithologic description of the core sequence from the Brooks Expl. Inc., Firestien No. 1-30 well, (2) sedimentological interpretation of that depositional sequence, and (3) calibration of the associated wire-line logs. After much of the study was completed, a second core sequence (the Brooks Expl. Inc., Leffler No. 1 core) was obtained, analyzed, and used to calibrate the associated wire-line logs. The following text presents the basic information derived from the analysis of the two core sequences presented in this study.

Lithologic Description of Core Sequence (Brooks Expl. Inc., Firestien No. 1-30)

Lithologic Unit	Lithologic Description	Thickness (ft)
1	Core depth 4240.0–4271.3 ft (approx. log depth 4243.7–4275.0 ft), including interval of missing core (4251.8–4265.2 ft): Burrowed to bioturbated mixture of muddy	

302

sandstone and sandy mudstone ("shale") with remnant laminated and ripple-bedded sandstone layers; very light gray (N7), very fine-grained sandstone layers biomixed with interlayered, medium to medium-light gray (N5-N4) silty mudstone; 5%-10% black carbonaceous grains and traces of glauconite and mica; poorly sorted overall; moderately indurated with detrital caly matrix and minor calcite cement; mainly burrow mottled to completely bioturbated, including a few distinct burrow types (e.g., *Planolites*, *Thalassinoides-Asterosoma*, minor vertical *Skolithos*, very small silt-lined burrows, *Chondrites*, and very small polychaete burrows); thin remnant laminated to ripple-bedded sandstone layers; also several laminated to ripple-bedded sandstone layers with sharp bases and burrowed tops: Core Laboratory analyses (N = 18) from this interval range in porosity from 10.4% to 16.6% (avg. = 14.6%), horizontal permeability from 0.03 to 0.95 md, and vertical permeability from 0.01 to 0.35 md; lower contact is sharp but relatively conformable.

31.3

2   Core depth 4271.3-4289.7 ft (approx. log depth 4275.0-4294.0 ft): Ripple-bedded and laminated and burrowed sandstone with "tight" calcite-cemented sandstone zones in lowermost and uppermost parts; quartzose sandstone with moderate amounts (5%-10%) of black carbonaceous debris, greenish glauconite (2%-5%), disseminated pyrite (1%-2%); well- to moderately well-sorted; very fine-grained; lightly calcite-cemented throughout most of interval except for intense cementation in lower 2-3 ft (4287.4-4289.7 ft) and upper 1.1 ft (4271.3-4272.4 ft); color very light gray (N8) to light gray (N7); sedimentary structures dominated by ripple bedding with thin clay laminae on ripple surfaces and plane bed lamination; burrow structures (mostly *Planolites*-like) are common throughout but do not completely destroy physical sedimentary structures; note upper 0.25 ft (4271.3-4271.55 ft) with well-rounded pebbles composed of calcareous pelletal mud mixed with quartzose sand: Core Laboratory analyses from the non-calcite-cemented intervals (N = 14) range in porosity from 13.1% to 20.2% (avg. = 17.3%), horizontal permeability from 0.57 to 22 md, and vertical permeability from 0.01 to 5.70 md; calcite-cemented zones range in porosity from only 3.0% to 10.4% (avg. = 5.2% for five samples) and permeability from only less than 0.01 to 0.03 md; lower contact appears to be very sharp, although lower 1-2 in. are not preserved in the core.

18.4

3   Core depth 4289.7-4300.0 (approx. log depth 4294.0-4304.3 ft): Bioturbated sandy mudstone; very poorly sorted, sandy and clayey siltstone to silty and clayey sandstone with 60%-70% clayey and silty matrix and 30%-40% very fine sand; color medium grey (N5); quartzose sand and silt and detrital clay; greenish glauconite (2%-5%), disseminated

pyrite (2%-4%), black carbonaceous debris (1%-5%), and white plagioclase(?) grains (5%-15%); moderately indurated; completely bioturbated with few distinct burrow structures (e.g., *Planolites*, silt-lined burrows): Core Laboratory analyses (N = 10) from this unit range in porosity from 10.1% to 11.7% (avg. = 10.8%) and permeability from less than 0.01 to 0.03 md; lower contact not present in core.

10.3

Lithologic Facies in the Firestien No. 1-30 Core

Three lithologic and sedimentologic facies can be delineated in the core Firestien No. 1-30 sequence of the Terry Sandstone. The lower part of the core displays the (F-1) *bioturbated mudstone* facies of the Pierre Shale that occurs below, and probably above, the Terry Sandstone throughout the study area. The (F-2) *ripple-bedded sandstone* facies of the lower part of the Terry Sandstone represents the best reservoir interval and displays a sharp, possibly erosional, basal contact. The ripple-bedded sandstone facies is overlain, with a rather short-interval transitional contact, by the (F-3) *burrowed to bioturbated muddy sandstone/sandy mudstone* facies of the upper part of the Terry Sandstone core sequence.

Facies F-1: bioturbated mudstone. Bioturbated sandy mudstone of the lower part of the core (core depth 4289.7-4300.0 ft) probably represents the marine "shelf" sediments present below the Terry Sandstone throughout the study area. The highly bioturbated medium to medium-dark gray sandy mudstone contains moderate amounts of fine, carbonaceous debris, greenish glauconite, and disseminated pyrite, and apparently represents deposition in a relatively quiet but well-oxygenated, marine setting that was most likely on the shelf at below wave-base depths. The mudstone has a very uniform lithologic character, as revealed in the wire-line logs. A thin marker (probably bentonite) occurs within this facies at about 30-40 ft below the Terry Sandstone; however, it was not present in the core for examination.

Facies F-2: ripple-bedded sandstone. The reservoir sandstone of the Terry Sandstone is represented by facies F-2, which appears to abruptly overlie facies F-1 with a sharp, and possibly erosional, contact at core depth 4289.7 ft. The exact nature of that contact could not be determined because the one or two critical inches had been removed prior to examination of the core (Fig. 6). Sandstone of this facies extends upward to core depth 4271.3 ft and can be characterized in general as being very fine grained, mostly moderately well sorted, and dominated by small-scale ripple bedding. Some plane-bed lamination and small-scale cross-bedding occur, and burrows (mostly, random nondescript *Planolites* structures) are common throughout. The sandstone is mainly composed of quartz, but contains significant amounts of fine, black carbonaceous debris and greenish glauconite and is moderately well indurated with moderate amounts of calcite and probable clay cements. Two intensely calcite-cemented intervals are present at the top and bottom of this facies, and they can be easily delineated as resistant, "tight" intervals on the wire-line logs (Fig. 4).

Facies F-2 probably represents relatively "high energy" transport and deposition of sand in a shallow marine environment. The predominance of ripple bedding with thin intervals of plane beds and small-scale cross beds obviously represents relatively strong currents. The occurrence of glauconite and presence of burrows throughout the sequence suggests a marine setting that must have been relatively shallow to have had steady to episodic strong currents. The sharp basal contact suggests that some erosion of the underlying sandy mudstone might have occurred prior to deposition of the basal sand.

The "tight" calcite-cemented zones obviously are secondary diagenetic features that seriously limit the vertical extent of the productive reservoir interval. Such intervals, however, probably only represent discontinuous calcite-cemented sandstone concretions that would only locally affect the reservoir.

Facies F-3: burrowed to bioturbated muddy sandstone/sandy mudstone. The ripple-bedded sandstone facies is overlain conformably, with transitional contact, by the biomixed interval of muddy sandstone and sandy mudstone ("shale") of facies F-3. The burrow structures that dominate this facies have almost completely destroyed the primary physical sedimentary structures, which were ripple-bedded and plane-bed laminated sandstone layers interbedded with laminated to wavy-bedded sandy mudstone. Besides the general burrow mottling, a few distinct burrow structures can be delineated, including *Planolites, Skolithos, Thalassinoides-Asterosoma, Chondrites*, silt-lined burrows (*Terebelina*), and tiny polychaete feeding burrows. Significantly, a few laminated to ripple-bedded sandstone layers do occur within this sequence, and they commonly display sharp bases and burrowed tops.

Facies F-3 includes both parts of the upper part of the lower sandstone unit and the lower part of the upper sandy mudstone unit of the Terry Sandstone (Figs. 4, 5). That vertical interval obviously represents a gradual decrease in physical (wave and current) depositional energy and increase in biogenic activity over that of facies F-2. Such a change was due perhaps to a gradual increase in water depth over the area. The few thin, relatively nonburrowed, laminated to ripple-bedded sandstone layers within the facies F-3 sequence probably represent periodic storm events in the marine shelf setting.

Discussion of Genetic Implications

The general basinal depositional setting for the Terry Sandstone is probably represented best by facies F-1 deposits, which comprise much of the Pierre Shale sequence in the basin. These sediments suggest open marine-shelf sandy mud deposition and associated intense bioturbation. The overlying Terry Sandstone's (1) sharp, probably scoured base, (2) good sand development in the lower part (facies F-2), and (3) distinct fining-upward sequence (facies F-3) suggests channelization and active to inactive channel fill. The sandstone to mudstone sequence of the Terry Sandstone throughout the Firestien No. 1-30 core displays a rather uniform mix of physical and biogenic sedimentary structures, which indicates relatively continuous, nonepisodic, deposition in a shallow marine setting.

Two main problems arise with the genetic implications indicated above: (1) the improbable occurrence of a subaqueous marine-shelf channel and (2) transport of sand across a marine shelf, which also could include areas of sediment bypass. These problems are being addressed in the ongoing study of the Terry Sandstone within the Denver basin. Detailed mapping of the sandstone body from which the Brooks Expl. Co., Firestien No. 1-30 core was obtained is presented later in this paper.

Lithologic Description of Core Sequence
(Brooks Expl. Inc., Leffler No. 1)

Lithologic Unit	Lithologic Description	Thickness (ft)

1   Core depth 4380.3-4381.5 ft (approx. log depth 4393.3-4394.5 ft): Cross-bedded and ripple-bedded sandstone with very thin clayey laminae; quartzose with minor to moderate amounts of sand-size claystone clasts (5%-7%), black carbonaceous debris (1%-3%) and greenish glauconite (tr-2%); well-sorted; fine-grained; moderately indurated with quartz, clay, and minor calcite cement; very light to light gray (N7-N8); sedimentary structures mainly small-scale cross bedding, ripple bedding, and wavy bedding with bedding surfaces delineated by fine laminae of clay and carbonaceous debris; minor *Planolites* burrows in lower part; measured porosity 14.9% and permeability 1.4 md; lower contact sharp and conformable.

                                              1.2

2   Core depth 4381.5-4382.0 ft (approx. log depth 4394.5-4395.0 ft): Laminated to ripple-bedded, calcite-cemented sandstone; quartzose with moderate amounts of black carbonaceous debris (1%-2%) and greenish glauconite (tr-2%); well-sorted; fine-grained; extremely well indurated with calcite cement (probably calcite-cemented sandstone concretion zone); color medium to medium-light gray (N5-N6); wavy lamination to minor ripple bedding; measured porosity 7.7% and permeability 0.1 md; lower contact not observed in core.

                                               0.5

3   Core depth 4382.0-4395.1 ft (approx. log depth), 4395.0-4408.1 ft): Laminated to ripple-bedded and burrowed sandstone with minor to moderate amounts of interlayered muddy and carbonaceous debris; quartzose with moderate amounts of sand-size mudstone clasts (2%-7%), black carbonaceous debris (1%-3%) and greenish glauconite (2%-4%), very fine- to fine-grained; well-sorted; minor *Inoceramus* shell fragments; moderately indurated with quartz, clay, and minor calcite cements; color very light to light gray (N7-N8) with laminae of black carbonaceous debris; sedimentary structures wavy lamination to ripple bedding and moderate amounts of *Planolites* and *Skolithos* burrow structures (general upward decrease in burrows although *Skolithos* burrows are better developed toward top of unit); good *Asterosoma* burrow structure at core depth 9394.4 ft; measured porosity 14.9%-18.9% (avg. = 17.1%; N = 13) and permeability 1.5-10.0 md (N = 13); lower contact transitional.

                                              13.1

4   Core depth 4395.1-4397.0 ft (approx. log depth 4408.1-4410.0 ft): Highly burrowed, muddy sandstone to sandy

mudstone; clayey and quartzose with moderate amounts of black carbonaceous debris (1%-4%) and greenish glauconite (2%-5%); poorly sorted, silt-size; color medium gray (N-5); moderately indurated with clay and quartz cements; sedimentary structures mainly burrow-mottled laminations and numerous burrow types (including *Skolithos, Planolites, Thalassinoides*, and *Chondrites*); measured porosity 13.1%-14.4% (N = 2) and permeability 0.1-6.5 md (N = 2); lower contact sharp and conformable.

1.9

5   Core depth 4397.0-4400.1 ft (approx. log depth 4310.0-4313.1 ft): Slightly "shaley" (i.e., muddy), laminated to ripple-bedded, and burrowed sandstone; quartzose and clayey with moderate amounts of sand-size mudstone clasts (2%-7%), black carbonaceous debris (1%-3%), and greenish glauconite (2%-5%); moderate sorting; very fine-grained; light gray (N-7); moderate induration with clay, quartz, and minor calcite cements; sedimentary structures mainly wavy lamination to ripple bedding and moderate abundance of *Thalassinoides* and *Planolites* burrows; porosity estimated at 12%-15% and permeability estimated at 2-5 md; lower contact not observed in core.

3.1

6   Core depth 4400.1-4402.1 ft (approx. log depth 4413-4415.1 ft): Laminated to ripple-bedded and burrowed, highly calcite-cemented sandstone; quartzose with moderate amounts of greenish glauconite (2%-5%) and extremely well indurated with calcite cement (approx. 35%-45% rock); moderately sorted; very fine-grained; medium to medium-light gray (N5-N6); sedimentary structures mainly wavy lamination to ripple bedding and moderate *Planolites* burrows; extremely low estimated porosity and permeability owing to intense calcite cementation; lower contact not observed in core.

2.0

7   Core depth 4402.1-4405.3 ft (approx. log depth 4415.1-4418.3 ft): Laminated to ripple-bedded sandstone with minor to moderate amount of "shale" (i.e., mudstone) laminae; quartzose with minor to moderate amount of detrital clay (5%-15%), black carbonaceous debris (2%-4%) and greenish glauconite (2%-5%); moderate to good sorting; very fine-grained; moderate induration with clay, quartz, and minor calcite cements; light gray (N7) with dark gray to black laminae; sedimentary structures mainly wavy lamination to ripple bedding and *Planolites* burrows; note distinct silt-lined plural-curving-tube structures and *Asterosoma* burrow structures; porosity estimated at 10%-15% and permeability estimated at 1.5-4.0 md; lower contact conformable and transitional.

3.2

8   Core depth 4405.3-4410.5 ft (approx. log depth 4418.3-4423.5 ft): Laminated and wavy-bedded to ripple-bedded

and burrowed sandstone with abundant interlayered sandy mudstone; quartzose and clayey with abundant sand-size mudstone clasts (5%-10%), black carbonaceous debris (1%-3%) and greenish glauconite (2%-4%); moderate sorting; silt to very fine-grained sand with interlayered mudstone; very light gray (N8) to medium gray (N5); moderately indurated with clay, quartz, and minor calcite cements; sedimentary structures mainly wavy lamination to ripple bedding with abundant *Planolites* and *Skolithos* burrows, as well as a few distinct *Ophiomorpha* and silt-lined plural-curving-tube structures; basal contact transitional.

5.2

9   Core depth 4410.5-4412.3 ft (approx. log depth 4423.5-4425.3 ft): Laminated to ripple-bedded and burrowed muddy sandstone with moderate amount of interlayered mudstone; quartzose with moderate amounts of detrital clay (5%-10%), sand-size mudstone clasts (5%-10%), black carbonaceous debris (2%-3%) and greenish glauconite (2%-4%); light to very light gray (N7-N8) with medium gray (N5) mud laminae; moderate induration with clay, quartz, and minor calcite cements; very fine-grained sand; moderate sorting; sedimentary structures mainly wavy lamination to ripple bedding and moderate *Planolites* burrows; lower contact not present in core.

1.8

Lithologic Facies in the Leffler No. 1 Core

The lithologic sequence displayed in the Leffler No. 1 core occurs entirely within the lower part of the Terry Sandstone Member. In general, three lithologic and sedimentologic facies can be delineated in the Leffler No. 1 core sequence. Some of the facies units are similar lithologically to those delineated in the Firestien No. 1-30 core; however, the overall character of the Leffler No. 1 facies and their position within the vertical sequence have influenced the decision to designate these additional facies indicated herein. The lower half of the core displays the (L-1) *burrowed to bioturbated sandstone/sandy mudstone* facies that appears to grade downward into the underlying Pierre Shale (i.e., facies F-1 of the Firestien No. 1-30 core). The (L-2) *ripple-bedded to burrowed sandstone* facies of most of the upper half of the Leffler No. 1 core is overlain by a thin interval (i.e., 1.2 ft) of

310

the (L-3) *ripple-bedded to cross-bedded sandstone* facies of the upper-most part of the coarsening-upward sandstone body of the lower part of the Terry Sandstone Member in the study area.

Facies L-1: burrowed to bioturbated sandstone/sandy mudstone. The lower six lithologic units of the Leffler No. 1 core sequence (core depth 4395.1-4412.3 ft) can be categorized as basal facies of the Terry Sandstone at that locality. Although the Pierre Shale and its contact with the Terry Sandstone are not expressed in the core, it is obvious from the wire-line log sequence (Fig. 7) that facies L-1 has a grada-tional basal contact with the underlying "shale" (i.e., facies F-1 of the Firestien No. 1-30 core). Facies L-1 is characterized by an under-layered sequence of highly burrowed to bioturbated sandstone, muddy sandstone, and sandy mudstone. Sand content increases gradually upward. Sedimentary structures are mainly wavy lamination and ripple bedding with abundant *Planolites, Thalassinoides*, and *Skolithos* burrows that, in some zones, completely obliterate the primary physical sedimentary structures. Clusters of silt-lined plural-curving-tube structures and a few distinct *Ophiomorpha* and *Asterosoma* structures can be delineated. Black carbonaceous wood and plant debris and glauconite are common compositional elements of this facies. An intensely calcite-cemented sandstone bed at core depth 4400.1-4402.1 ft (approx. log depth 4413.1-4415.1 ft) is well displayed as a high-resistivity "kick" on the induction log (Fig. 7). An overlying sandy mudstone unit at core depth 4395.1-4397.0 ft (approx. log depth 4408.1-4410.0 ft) can be easily delineated on the gamma-ray log (Fig. 7).

Facies L-2: ripple-bedded to burrowed sandstone. Most of the upper part of the Leffler No. 1 core (i.e., core depth 4381.5-4395.1 ft) is composed of laminated to ripple-bedded and burrowed sandstone with minor amounts of interlayered muddy and carbonaceous debris. This quartzose and glauconitic (2%-5%) sandstone facies displays relatively good reservoir properties (i.e., porosity of up to 19% and permeability of up to 10 md). Except for the intensely calcite-cemented bed at the top of the unit, the well-sorted and fine-grained sandstone is only moderately indurated with quartz (overgrowths), clay, and minor calcite cements. Sedimentary structures are mainly wavy lamination and ripple bedding with scattered burrows. *Planolites* structures are present throughout. Vertical *Skolithos*-like structures are common and increase in abundance upward.

Facies L-3: ripple-bedded to cross-bedded sandstone. The upper 1.2 ft of the Leffler No. 1 core (i.e., core depth 4380.3-4381.5 ft) is cross-bedded to ripple-bedded, fine-grained sandstone that appears to be representative of the upper 4.5 ft of the coarsening-upward sandstone of the Terry Sandstone at this locality (i.e., log depth 4390.0-4394.5 ft). This unit is characterized by the dominance of physical sedimentary structures and general lack of biogenic sedimentary structures. The moderately indurated sandstone is cemented with quartz (overgrowths), clay, and minor calcite, and displays relatively good reservoir properties.

Discussion of Genetic Implications

The core and wire-line log sequence of most of the lower part of the Terry Sandstone in the Brooks Expl. Inc., Leffler No. 1 well obviously is indicative of the conventional coarsening-upward offshore marine-shelf sandbar model. Such a sequence is in strong contrast with that of the fining-upward sequence displayed in the Brooks Expl. Inc., Firestien No. 1-30 well. Facies L-1 overlies the Pierre Shale with an apparent transitional contact, coarsens upward, and is overlain sharply by the better-developed sandstone of facies L-2 and L-3. The burrowed-to-bioturbated interlayed sandstone and mudstone sequence of facies L-1 is indicative of relatively continuous, nonepisodic deposition in a shallow marine setting. Sandstones of facies L-2 and L-3 become coarser, better sorted, and increasingly dominated by physical sedimentary structures in an upward direction in the core. Such a sequence indicates an increase in current energy, possibly as a result of marine sandbar buildup and associated decrease in water depth.

The main part of the lower unit of the Terry Sandstone Member in the Leffler No. 1 well represents an upward-coarsening offshore marine sandbar deposit. Of interest is the presence of an overlying shale bed 1.5 to 2.0 ft thick (approx. log depth 4388-4390 ft) and a thin sandstone unit (approx. log depth 4383-4388 ft) at the top of the lower unit of the Terry Sandstone Member. The distinct shale bed and overlying 5-ft sandstone unit may be considerably different genetically than the coarsening-upward sandstone unit below. Mr. Bill Brooks (Brooks Exploration) reported that the uppermost sandstone unit gave a good show of oil, although producing water. The indication of hydrocarbons led him to develop the main oil reservoir unit of the La Poudre South Field area, from which the Firestien No. 1-30 core sequence was obtained (pers. comm., March 1986). The following text reveals how these sandstone bodies of the lower part of the Terry Sandstone Member are related.

Sandstone Body Geometry and Stratigraphic Relationships

In order to better understand the genesis of the Terry Sandstone sequence delineated in the Brooks Expl. Inc., Firestien No. 1-30 core, we obtained subsurface logs for most of the wells within the area shown in Figure 3, and a series of subsurface maps and cross sections were prepared to delineate sandstone body trend and geometry. Lithologically useful gamma-ray and resistivity logs were obtained for most of the 203 well locations used. The Brooks Expl. Inc., Leffler No. 1 core sequence was obtained at a late stage of the investigation in order to provide additional core material for the core workshop and to better document the character of the Terry Sandstone sequence in the study area.

Initial examination of the wire-line logs revealed three significantly different log patterns (Fig. 10). Subsequent generation of subsurface maps and cross sections revealed the presence of two northwest-trending, elongate, and lens-shaped sandstone bodies that (1) are significantly different in terms of geometry, stratigraphic relationships, and origin; (2) display an overlapping, shinglelike lateral and vertical geometric relationship; and (3) are separated by an impermeable shale-like barrier.

Vertical Sequences and Terry Sandstone Distribution

The Terry Sandstone Member within the study area thickens from slightly less that 60 ft in the northwest to greater than 120 ft within 11 mi to the southeast (Fig. 3). The sequence can be delineated relatively

314

easily from the silty marine mudstone of the Pierre Shale below and above, although log picks of the lower or upper contact can be difficult where the contact is obviously transitional (Fig. 10). The sequence also can be rather easily subdivided into a lower sandstone unit and an upper sandy mudstone unit. The upper sandy mudstone unit forms a relatively uniform blanketlike sequence that ranges from about 50 to 70 ft thick throughout the study area and overlies the lower, more variable sandstone unit (see Figs. 10, 11, 13, 16).

The lower sandstone unit of the Terry Sandstone displays considerable variability across the study area. The three "end member" log patterns shown in Figure 10 display the vertical sequences of the two distinct sandstone bodies present and the general nature of the vertical sequence within the area of overlap of those sandstone bodies. The lower sandstone unit of the Terry Sandstone in the northeastern part of the study area is characterized by a gradational base, coarsening-upward log pattern, and sharp upper contact with the overlying sandy mudstone unit. In the southwestern part of the study area, the lower sandstone unit displays a sharp basal contact, fining-upward log pattern, and gradational contact with the overlying sandy mudstone unit. These two sandstone bodies display an overlapping relationship across a strip 1–2 mi wide that trends northwestward across the study area (Fig. 3). The wire-line log pattern of the lower sandstone unit of the Terry Sandstone within that area of overlap is characterized by a distinct shale "notch" that separates the two sandstone bodies (Fig. 10).

Sedimentary Bodies of the Lower Part of the Terry Sandstone

The lower part of the Terry Sandstone is composed of three major northwest-trending, elongate, lens-shaped sedimentary bodies, delineated herein as (1) northeastern sandstone body, (2) southwestern sandstone body, and (3) shale "notch." Wire-line log cross section A-A' (Fig. 11) displays the significant stratigraphic relationships of the three sedimentary bodies in a direction approximately parallel to depositional dip. The northeastern sandstone body is overlain along its southwestern flank by the shale notch unit, which extends from near the top of the sandstone body down its flank to the basal pinch-out along its southwestern margin. The shale notch unit is in turn overlain by the southwestern sandstone body in an overlapping manner. Therefore, the shale notch serves not only to help delineate the two major sandstone bodies of the study area, but also acts as an effective permeability barrier between the two bodies.

Northeastern sandstone body. The sandstone body of the lower part of the Terry Sandstone sequence in the northeastern part of the study area is displayed in Figures 12 and 13 and can be characterized according to the following important features: (1) northwest-trending, southeastward thickening, elongate geometry; (2) relatively symmetrical, lens-shaped cross section having a flat base and convex-upward top; (3) maximum thickness of 70 ft and maximum width of 4-5 mi with sharp depositional pinch-out along southwestern margin and gradational "silting out" along the northeastern margin; (4) vertical sequence with gradational lower contact, coarsening-upward textural trend, and sharp upper contact. Other significant features shown in Figure 13 include (5) a siltstone "cap," up to 10 ft thick, which thickens in the same direction that the sandstone body thickens, and (6) probable lack of a basal scour, as evidenced by the interval between the base of the sandstone and a bentonite marker bed approximately 10 ft below; that interval actually increases slightly along with the southeastward thickening of the sandstone body.

The Brooks Expl. Inc., Leffler No. 1 core sequence is representative of the general character of the vertical sequence of the northeastern sandstone body. This "typical" Cretaceous marine-shelf type sandstone unit grades upward from a highly burrowed to bioturbated sandy mudstone and muddy sandstone in the lower part to a burrowed, laminated, ripple-bedded and even cross-bedded, very fine- to fine-grained sandstone in the upper part. The sandstone unit displays a sharp upper contact with overlying mudstone. Borehole tests of the Terry Sandstone interval in a few of the wells through the northeastern sandstone body show only the presence of noncommercial volumes of gaseous hydrocarbon and some oil.

Shale "notch." The shale notch sedimentary body is named from the notchlike character of the shale unit observed in the gamma-ray/ resistivity log suites (see Fig. 10). The dip-oriented wire-line log cross section A-A' (Fig. 11) illustrates how the shale notch appears to "migrate" in a southwestward direction from the top to the bottom of the lower sandstone unit of the Terry Sandstone sequence. On the basis of the high-gamma/low-resistivity wire-line log character, the unit is inferred to be a uniformly low-silt-content clay shale. Duane Moredock (pers. comm., Feb. 1986) has recognized similar shale units within the Terry Sandstone of the Spindle Field area that, when cored, revealed a dark gray to black "poker-chip" shale. Louise Kiteley (pers. comm., Feb. 1986) indicated that where she has observed this unit in core, it is a "marine reworked volcanic tuff."

The shale notch forms a northwest-trending, elongate sedimentary body with a maximum width of 1.5 mi and maximum thickness of approximately 12 ft (Fig. 14). This shale body is slightly asymmetrical in cross section, having a more rapid thinning trend toward the southwestern margin. Of considerable significance is the manner in which the shale unit acts as a permeability barrier to vertically and laterally separate the two sandstone bodies of the lower sandstone unit of the Terry Sandstone. The origin of the sedimentary body is presently unknown and cannot be easily speculated on until additional core material can be examined in greater detail.

Southwestern sandstone body. The sandstone body of the lower part of the Terry Sandstone sequence in the southwestern part of the study area is displayed in Figures 15 and 16 and has the following important features: (1) northwest-trending, southeastward-thickening, elongate geometery; (2) strongly asymmetical lens-shaped cross section with concave-upward base, relatively flat top, and more rapid thinning trend toward the northeast where the sandstone body laps up onto the shale notch (and where the shale notch overlies the southwestern flank of the northeastern sandstone body); (3) maximum thickness of 50 ft and maximum width of 5-6 mi with relatively rapid pinch-out along the northeastern margin and very gradual "silting out" along the southwestern margin; and (4) vertical sequence with sharp lower contact, finingupward textural trend, and gradational upper contact. Other significant features shown in Figure 15 include (5) development of a southeastward-thickening, basal sand-rich unit that also appears to be best developed along the northeastern flank of the sandstone body and (6) probable basal scour, as evidenced by the interval between the

318

base of the sandstone body and a bentonite marker bed 10-30 ft below; that interval thickens significantly in the same southeasterly direction, and about the same amount, as the sandstone body.

The Brooks Expl. Inc., Firestien No. 1-30 core sequence displayed in this paper is representative of the vertical sequence of the southwestern sandstone body. This sandstone body also is oil productive in several sections of the study area (Fig. 15).

Discussion and Conclusions

The Terry Sandstone stratigraphic sequence of the middle part of the Upper Cretaceous Pierre Shale in the Denver basin of northeastern Colorado has been interpreted in general as a marine-shelf depositional sequence. Within the two-township area of Weld County delineated in this paper, the lower sandstone unit of the Terry Sandstone is composed of two distinct, northwest-trending, elongate, lens-shaped sandstone bodies. The sandstone bodies have been designated herein as the "northeastern" and "southwestern" sandstone bodies. They display a shinglelike overlapping relationship across a 1- to 2-mi wide, northwest-trending strip and are separated by a permeability barrier unit designated herein as the shale notch.

The northeastern sandstone body displays most of the geometric and wire-line log characteristics of the conventional offshore marine sandbar, including a rather symmetrical lens-shaped cross section, gradational base, coarsening-upward textural sequence, and sharp upper contact with overlying marine mudstone. The southwestern sandstone body is more indicative of some sort of marine-influenced, shallow shelf(?), channel deposit, as suggested by its asymmetrical, lens-shaped cross section, sharp erosional base, fining-upward textural sequence, and gradational upper contact with overlying marine mudstone. The shale notch unit is a major curiosity in this study; however, it has great significance in terms of delineating the two sandstone bodies and as a permeability barrier between those bodies.

The following genetic sequence is presented as a preliminary interpretation of the lower sandstone unit of the Terry Sandstone Member:

1. The northeastern sandstone body developed as an offshore marine-shelf sandbar.

2. The shale notch was deposited as a blanket of organic-rich, silt-poor clay and preserved along the landward (southwestern) flank of the marine sand bar; influence of eustatic, and/or tectonic, sea level change, or the general position of the shoreline during formation of the shale notch deposit has not yet been evaluated.

3. The southwestern sandstone body formed as a marine-influenced, channel-like deposit that scoured out a few feet of the underlying Pierre Shale mud and part of the shale notch deposit. This channelized sandbody has an active channel-fill, sand-rich lower unit along its steeper northeastern margin that abuts the preexisting marine sandbar. The location of the channel-like deposit may have been strongly influenced by the preexisting marine sandbar as dip-oriented currents were diverted in a southeasterly direction along the landward side of the marine sandbar.

We inferred many of the genetic aspects of the Terry Sandstone sequence from the geological analysis of the Brooks Expl. Inc.,

Firestien No. 1-30 core sequence from the southwestern sandstone body. Physical and biogenic sedimentary structures, sediment texture and composition, and vertical sedimentological relationships provided valuable guidelines as to the physical, biological, and chemical processes that influenced formation of the sedimentary deposit. Subsequent calibration of wire-line logs and construction of subsurface wire-line log cross sections and maps provided additional significant information with which we generated a preliminary genetic model for the formation of the Terry Sandstone sequence. That model can now be evaluated, tested, and further modified using additional data gathered according to the same general investigative procedures. This approach can be classified in general as the process sedimentology approach to the interpretation of subsurface sedimentary sequences. It can be a very powerful subsurface exploration tool if applied objectively and systematically.

References

Hobson, J. P., Fowler, M. L., and Beaumont, E. A., 1982, Depositional and statistical exploration models, Upper Cretaceous off-shore sandstone complex, Sussex Member, House Creek Field, Wyoming: Amer. Assoc. Petrol. Geol., Bull., v. 66, no. 6, p. 689-707.

Kiteley, L. W., 1976, Marine shales and sandstones in the Upper Cretaceous Pierre Shale at the Francis Ranch, Laramie County, Wyoming: Mountain Geol., v. 13, no. 1, p. 1-19.

Kiteley, L. W., 1977, Shallow marine deposits in the Upper Cretaceous Pierre Shale in the northern Denver basin and their relation to hydrocarbon accumulation: Rocky Mountain Assoc. Geol., 1977 Symposium, p. 197-211.

Kiteley, L. W., 1978, Stratigraphic sections of Upper Cretaceous Pierre Shale in the northern Denver basin, northeastern Colorado and southeastern Wyoming: U.S. Geological Survey Oil and Gas Charts OC-78, sheets 1-3.

Moredock, D. E., and Williams, S. J., 1976, Upper Cretaceous Terry and Hygiene sandstone, Singletree, Spindle and Survey fields, Weld County, Colorado, in R. C., Epis, and Weimer, R. J., Studies in Colorado field geology: Colorado School of Mines Professional Contribution No. 8, p. 264-74.

Porter, K. W., 1976, Marine shelf model, Hygiene Member of Pierre Shale, Upper Cretaceous, Denver basin, Colorado, in Epis, R. C., and Weimer, R. J., eds., Studies in Colorado field geology: Colorado School of Mines Professional Contribution No. 8, p. 251-63.

Porter, K. W., and Weimer, R. J., 1983, Diagenetic sequence related to structural history and petroleum accumulation, Spindle Field, Colorado: Bull. Amer. Assoc. Petrol. Geol. Bull., v. 66, p. 2543-60.

Siemers, C. T., and Tillman, R. W., 1981, Recommendations for the proper handling of cores and sedimentological analysis of core sequences, in Siemers, C. T., Tillman, R. W., and Williamson, C. R., eds, Deep-water clastic sediments, a core workshop: Soc. Econ. Pal. Min. Core Workshop No. 2, San Francisco, May 30 and 31, 1981, p. 20-44.

Tillman, R. W., and Martinsen, R. S., 1984, The Shannon shelf-ridge sandstone complex, Salt Creek anticline area, Powder River basin, Wyoming, in Tillman, R. W., and Siemers, C. T., eds., Siliciclastic shelf sedimentation: Soc. Econ. Pal. Min. Spec. Publ. No. 34, p. 85-142.

Weimer, R. J., 1973, A guide to the uppermost Cretaceous stratigraphy, Central Front Range, Colorado--deltaic sedimentation, growth faulting and early Laramide crustal movement: Mountain Geol., v. 10, no. 3, p. 53-97.

LITHOFACIES, INFERRED PROCESSES, AND LOG RESPONSE CHARACTERISTICS OF SHELF AND SHOREFACE SANDSTONES, FERRON SANDSTONE, CENTRAL UTAH

S. L. Thompson,[1] C. R. Ossian,[2] and A. J. Scott[3]

[1]ARCO Oil and Gas Company, P.O. Box 1610, Midland, Texas 79702

[2]ARCO Oil and Gas Company, P.O. Box 2819, Dallas, Texas 75221

[3]RPI Texas, 5910 Courtyard Drive, Austin, Texas 78731

Abstract

Shelf and shoreface sequences form the lowermost sandstone deposits and the leading edge of a progradational deltaic wedge (Upper Cretaceous Ferron Sandstone Member of the Mancos Shale, Central Utah). Cores taken by ARCO Oil and Gas Company penetrated these shelf and shoreface sequences about a mile from their outcrop in Castle Valley, Utah. Outcrop evidence suggests that the shelf sandstone/-siltstone unit underlying the Ferron Delta complex is the distal part of a shelf sand body originating from another delta complex at least 50 mi to the north. Possible amalgamated shelf sand deposits have been recognized at the base of a shoreface sequence in one core. The Ferron cores offer excellent opportunities to examine wire-line log and core recognition criteria for shelf sandstone sequences in association with deltaic settings. Shelf sands associated with delta systems have been shown to be important hydrocarbon exploration targets.

Ferron Delta progradational facies consist of two main assemblages: bioturbated shelf siltstones and sandstones underlying the delta proper, and upward-coarsening delta front sandstones. The bioturbated shelf facies consist of several lithofacies which include laminated mudstone, rippled and burrowed siltstone, and bioturbated sandy siltstone. Component lithofacies of the upward-coarsening sandstone interval are rippled sandy siltstone, interbedded sandstone and siltstone, horizontally bedded sandstone, and cross-bedded sandstone.

Introduction

Features observed in the cores provide criteria for the recognition of shelf and shoreface sandstone facies. Processes inferred from sedimentary structures and characteristics of these facies have been integrated with regional surface and subsurface analyses of the Ferron Sandstone in central Utah (Fig. 1) to develop a depositional model.

These cores are part of an extensive database (Fig. 2) compiled by ARCO Oil and Gas Company geologists to determine stratigraphic relationships of the Ferron Sandstone (Thompson and Ossian, 1985). The excellent Ferron Delta exposures and the availability of extensive subsurface control (from the Ferron gas field and coal mines in the area) have long attracted the attention of stratigraphers and sedimentologists. This sandstone complex is associated with a major fluvial/deltaic wedge that prograded generally eastward into a rapidly subsiding foreland basin developed along the western margin of the Cretaceous Interior Seaway (Cotter, 1975a; Ryer and McPhillips, 1983; Thompson, 1985).

Objectives

Objectives of this contribution to the core workshop are to (1) document Ferron shelf sandstone characteristics, (2) discuss processes that influenced deposition, (3) integrate shelf sandstone facies into the regional deltaic depositional model, and (4) compare and contrast these shelf sandstones with associated Ferron deltaic shoreface facies.

Modern Shelf Processes and Deposits

In recent years, major advances have been made in understanding the effect of depositional processes on a variety of modern continental shelf deposits. Documentation of tidal current influences in the North Sea has provided a spectrum of tide-dominated shelf sand ridge types, while work in storm-dominated areas of the eastern North American

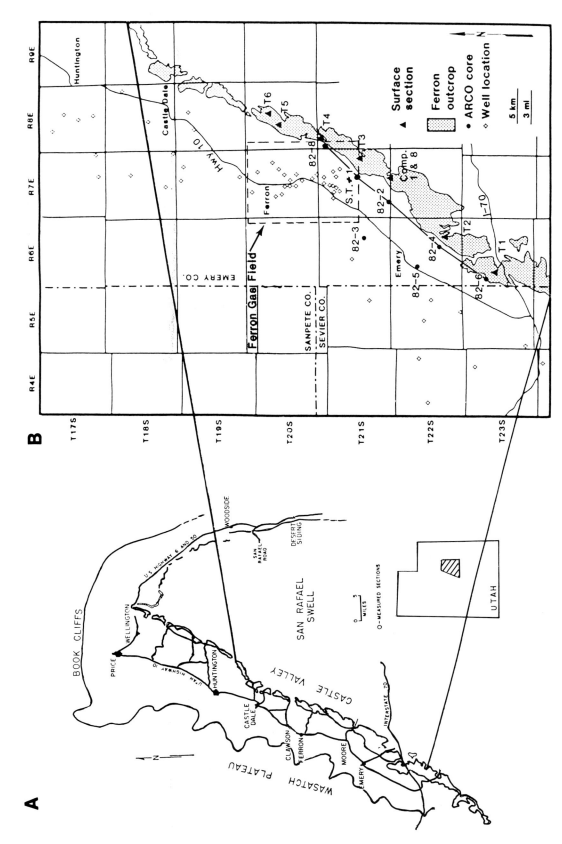

Figure 1. Castle Valley area. A. Exposure of Ferron Sandstone in Castle Valley (from Cotter, 1975). B. Location of cores and other data points used in Ferron study (Thompson, 1985).

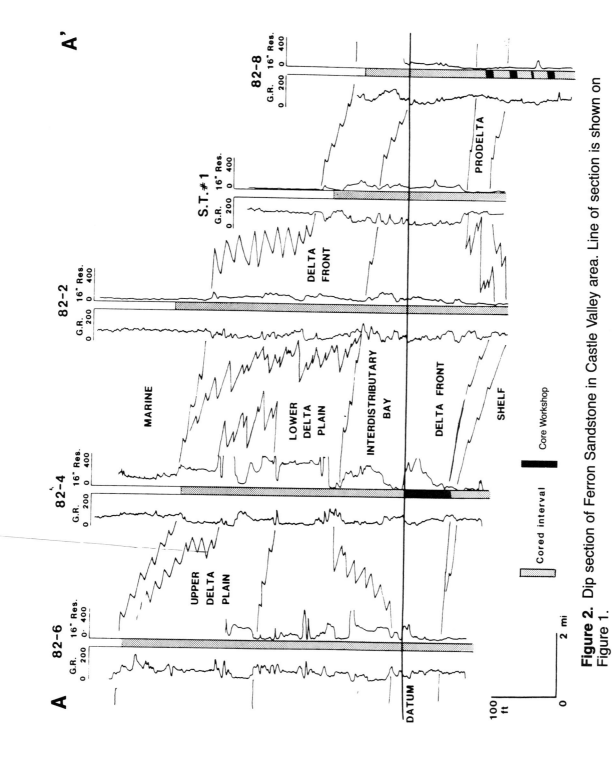

Figure 2. Dip section of Ferron Sandstone in Castle Valley area. Line of section is shown on Figure 1.

328

Atlantic shelf and on the Oregon-Washington shelf has also aided greatly in providing a basis for better understanding of ancient shelf deposits. It is beyond the scope and purpose of this paper to review this research, but an introduction to the complexities of modern shelf processes and depositional patterns is available in many fine publications (e.g., Reading, 1978; Nittrouer, 1981; Stubblefield et al., 1984; Swift et al., 1984). Shelf processes that generate sand bodies on the shelf near deltaic headlands have been described by Coleman et al. (1981).

The discovery of large sand bodies of reservoir quality located on the shelf seaward from the expected shoreline trend is highly important to "shelf exploration." The processes of sand emplacement and sediment sources are still controversial, but such sand bodies are now known to be present in the Utah Turonian shelf in Utah. We shall demonstrate recognition criteria for such sand bodies in this workshop.

Coleman et al. (1981) have described arcuate sand bodies in shelf settings near deltaic headlands of the modern Nile River. Relict shelf sand bodies (older distributary mouth bars deposited at lower sea level) are shaped and distributed by strong longshore currents deflected by the distributary mouth complex of the Damietta Branch of the Nile River (Fig. 3). The sand bodies acquire their arcuate shape by the combined effects of the deflected easterly longshore drift and strong westerly countercurrents. We believe that similar systems were at work in the Cretaceous Turonian Seaway of Utah.

Coleman (pers. comm.) referred us to similar shelf sand bodies associated with the modern Rhone and Ebro river deltas. In each of these areas, he and his co-workers ascribe these bodies to reworked

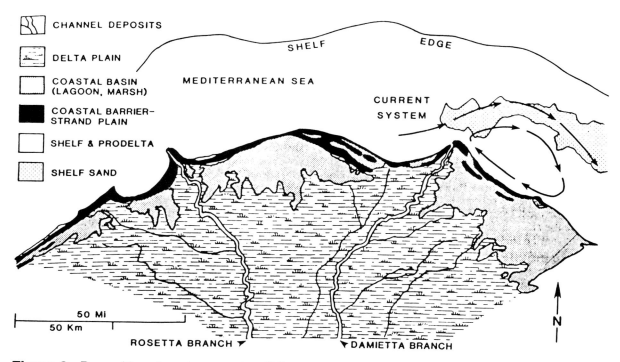

Figure 3. Depositional environments of Nile delta system and associated shelf (from Palmer and Scott, 1984).

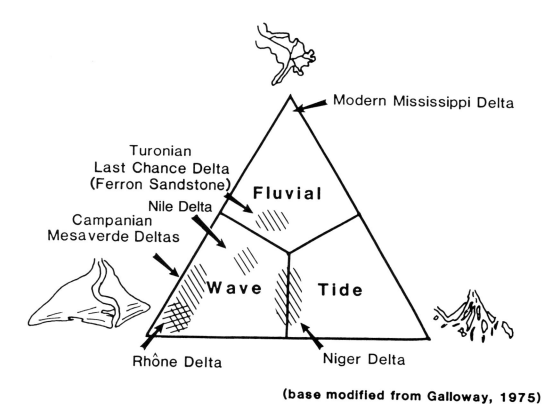

Figure 4. Classification of delta systems according to processes that influence deposition (after Galloway, 1975).

delta front systems established during earlier, lower sea-level lowstands.

Sand deposited on deltaic distributary channel mouth bars might be transported basinward onto the prodelta shelf by three major processes. In tide-dominated areas, sands from the distributary may be reworked by tidal currents. In other areas, major storms or floods may breach the channel mouth bar crest, eroding sediments and transporting them basinward by density underflows (turbidity currents) to be deposited on the distal bar slope or in adjacent prodelta areas. The resultant thin sand beds have been referred to as frontal splay deposits.

A third process that forms potentially larger-scale shelf sand deposits has been suggested by Palmer and Scott (1984) (Fig. 3). Coastal shelf currents may be diverted basinward near channel mouth bars or deltaic headlands. These diverted currents may transport sediments onto the shelf forming an arcuate "plume-shaped" deposit that extends downdrift and onto the adjacent muddy shelf (Palmer and Scott, 1984). These coarser clastic sand bodies are then stored in shelf areas beyond the influence of normal shallow shelf longshore energy systems. (The usage here of the word "plume' is an allusion to the shape of these sand bodies, i.e., an ostrich feather.)

Two major mechanisms appear to account for the preservation of these shelf sandstone deposits. A shift in a deltaic axis or a marine transgression may cause the abandonment of a shelf "plume." The sediments of the abandoned shelf sand bodies are then buried by shelf muds deposited from suspension. If sea level remains stable and deltaic progradation continues, these deposits will ultimately be overrun and amalgamated to the lower deltaic shoreface. Whether these shelf sand bodies are reworked from earlier delta front sands established at lower sea levels or were transported basinward from the distributary mouth bar remains subject to controversy.

Due to outcrop and subsurface limitations, it has not yet been possible to determine which of these scenarios might have produced the Utah arcuate sand body examples described below.

Ferron Sandstone

The Ferron Sandstone crops out for about 60 mi along the cliffs of Castle Valley, central Utah (Fig. 1). The sand-rich interval ranges in thickness from more than 500 ft in the coal-rich delta plain sequence in the southwest part of the outcrop to less than 100 ft in the delta front and shelf sequence to the northeast. A suite of five cores was taken by ARCO Oil and Gas Company along a line parallel to and about 1 mi from the outcrop. An additional two cores were taken along a parallel line approximately 3-4 mi from the outcrop.

Five of these cores are arranged along a cross section (Fig. 2) to form a dip section through a northeastward thinning clastic wedge. The interval consists mostly of shelf and delta front deposits in the northeast and a thick accumulation of delta plain deposits in the southwest. The thick deltaic accumulation resulted from crustal loading and downwarping in a foreland basin with greater subsidence along its western margin. The large sediment supply from the Cordillera filled the subsiding basin and continued to prograde eastward onto the shelf. When sediment supply was no longer able to remain in balance with subsidence, the delta foundered and a local marine transgression occurred.

Depositional Model

Detailed examination of more than 3000 ft of core led to recognition of several distinctive component facies based on suites of sedimentary structures, textures, and other features. These facies were also studied in outcrop, and their spatial and stratigraphic relationships determined. Our depositional model (Fig. 5) is based on these relationships and inferred depositional processes.

The Ferron Sandstone in Castle Valley was deposited as a lobate mixed-energy delta complex along the western margin of a shallow seaway in which wave, storm, and tidal processes were approximately balanced with riverine processes (Fig. 4). The dominant depositional processes were dependent upon relative position on the delta platform such that longshore currents and waves were dominant forces off the flanks of the delta. Fluvial processes had the greatest influence at the mouth of the distributaries, and tidal and storm processes had considerable impact on the associated lagoons and interdistributary bays. The result of these processes was a lobate delta (Fig. 5) with flanking barrier islands, backbarrier lagoons, and interdistributary bays with received crevasse splays and washover fans. A large stable delta plain developed, and delta lobes switched location frequently by channel avulsion as distributaries overextended their gradient advantage. Subsidence produced thick accumulations of delta plain deposits and stacking of delta front sandstones.

Two types of sandy shelf facies have been recognized. One facies (Fig. 5, facies A) is a very widespread sheet of fine sandstone and siltstone derived from the Vernal Delta about 50 mi to the north and reworked by shelf currents and bioturbating organisms. This

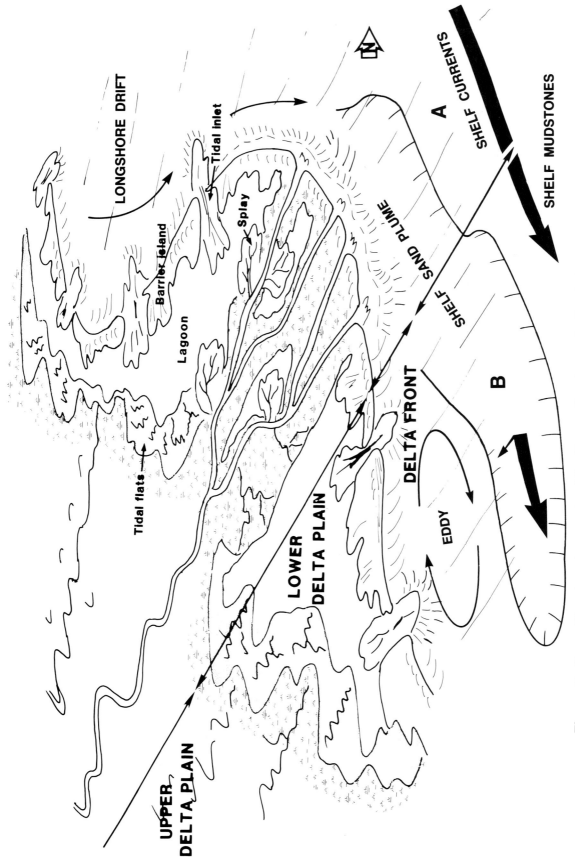

Figure 5. Depositional environments of the Ferron Sandstone in Castle Valley, Utah. Two types of shelf deposition are represented by (A) shelf mudstones transported from outside immediate area and (B) fine sandstones transported shelfward from advancing delta.

amalgamated set of shelf deposits is 80 ft thick in core 82-8, and outcrop observation indicates that it becomes sandier toward its source. Although the actual source of Ferron shelf sand bodies has not yet been determined, these shelf deposits may represent "plume-shaped" bodies derived from the nearby Vernal Delta complex.

The second shelf facies type (Fig. 5, facies B) is a much smaller scale feature. Several observed shoreface intervals are sandier at their bases than would be expected by normal progradational processes (as in core 82-4). These sandy lower shoreface accumulations may be amalgamated shelf "plume" sandstones. Such sediment "plumes" would have been adjacent to contemporaneous distributary mouth bars and later overrun by prograding deltaic deposits. This process would cause amalgamation with the advancing lower shoreface facies tract. Parts of cores 82-8 and 82-4 have been selected for the core workshop to illustrate these two distinct types of shelf sand sequences.

Figure 6A shows the outcrop extent and energy patterns of the shelf sand body derived from the Vernal Delta to the north. The unit consists of fine sandstone near its origin by Wellington, Utah, and gradually fines to scattered siltstone lentils east of Emery, Utah. Outcrops near Castledale, Utah, demonstrate the western edge of the main sand body and the margins of the eddy facies, while other sections near the sites of Cedar and Mounds, Utah, record the eastern edge of the main shelf sand body and its gradational merger with the nonclastic shelf facies.

ARCO cores 82-4 and 82-8 are located in the distal portions of the main "plume" (Fig. 6A). Other similar shelf sand bodies have been recognized in the Castle Valley area within the Ferron progradational

335

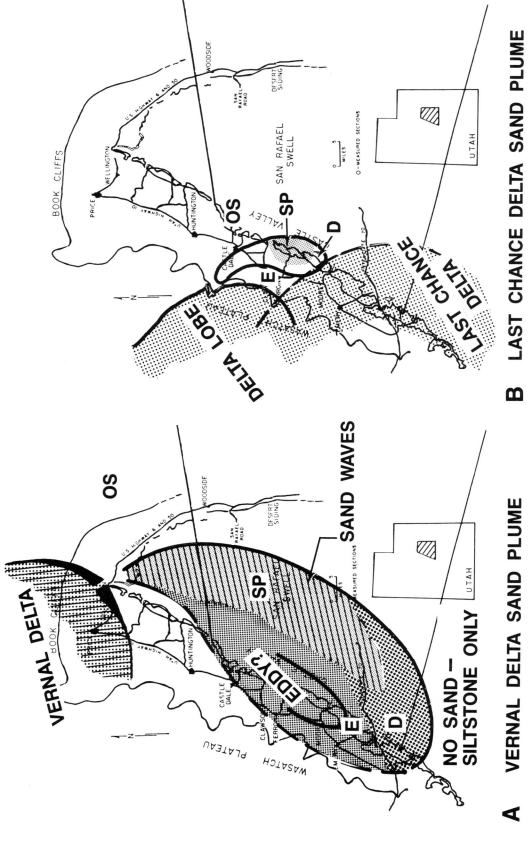

Figure 6. Shelf sand plumes of Castle Valley area. A. Vernal Delta sand plume; B. Possible plumes introduced into area from advancing deltaic systems. OS — open shelf; SP — sand plume; E — eddy; D — distal silty plume.

A   VERNAL DELTA SAND PLUME

B   LAST CHANCE DELTA SAND PLUME

interval. One of these with outcrop expression near Ferron, Utah (Fig. 6B) consists of a field of isolated unidirectional sand waves and represents the very distal margin of the sandier shelf body facies.

Core Descriptions

Cores and measured sections were described using a graphic logging procedure (Fig. 7). Grain size and sedimentary textures were determined using a grain-size comparator, while sedimentary structures and stratification types were sketched from the core. Graphic core descriptions are more easily compared to log response patterns or measured sections of outcrops than are other data recording methods. Graphic logging forms similar to those used here have been discussed by Boyles and Scott (1982).

ARCO Oil and Gas Company Core 82-8

ARCO 82-8 (Figs. 8, 9) is from the most distal part of the deltaic complex, thereby containing almost exclusively marine rocks. The main facies assemblage consists of laminated mudstone, overlain by a burrowed siltstone interval, bioturbated sandy siltstone, and silty sandstone. The intensive bioturbation and fine grain size suggest deposition several miles from the shoreline in somewhat deeper water (inner to middle neritic). In outcrop, this unit forms a bench which underlies the Ferron Sandstone and thickens northward (Cotter, 1975b) toward its apparent source. The coarsest sediment fraction observed throughout the area was coarse silt to very fine sand. This sequence differs from the Duffy Mountain Sandstone shelf sequence described by Boyles and Scott (1982) which is coarser grained, has well-defined

337

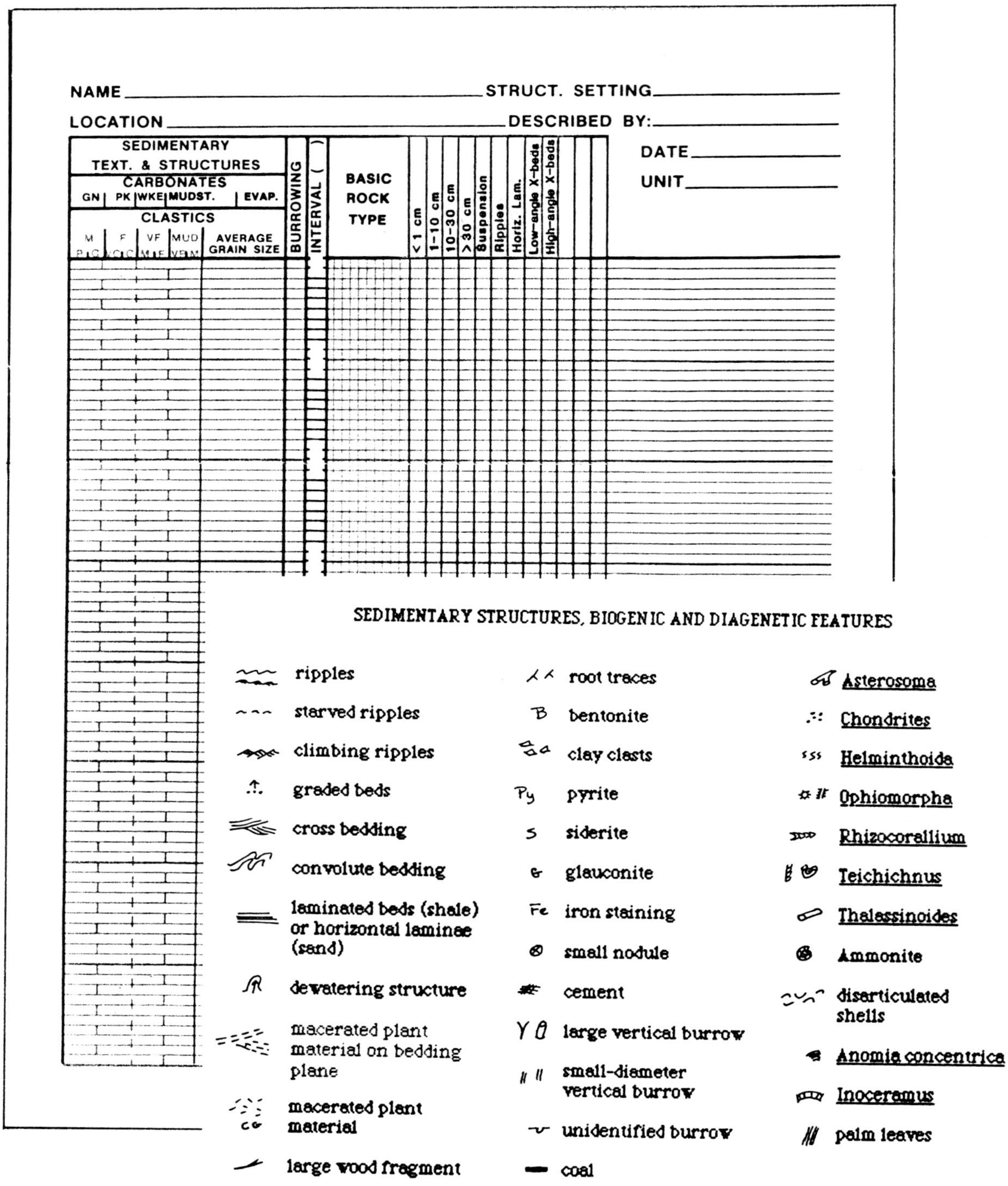

Figure 7. Core description form and symbols used.

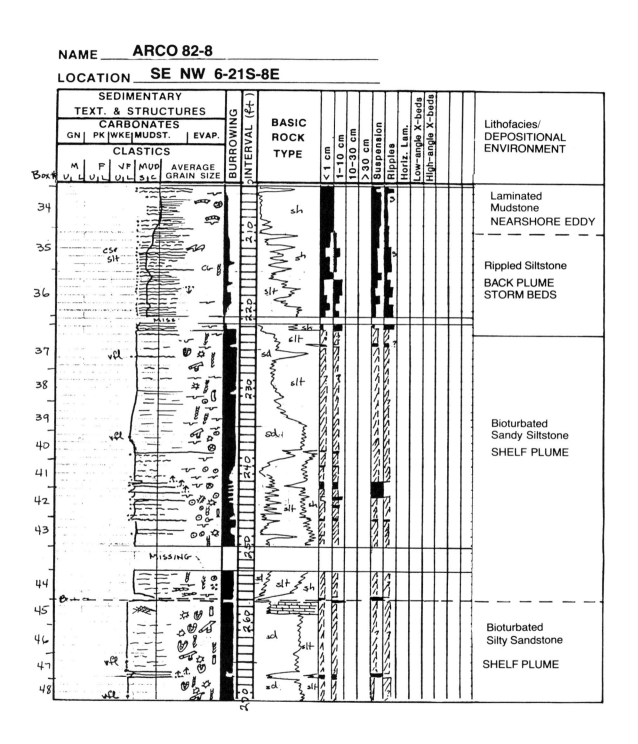

Figure 8. ARCO Core 82-8 description and lithofacies interpretation.

Figure 8. (cont.) ARCO Core 82-8 description and lithofacies interpretation.

341

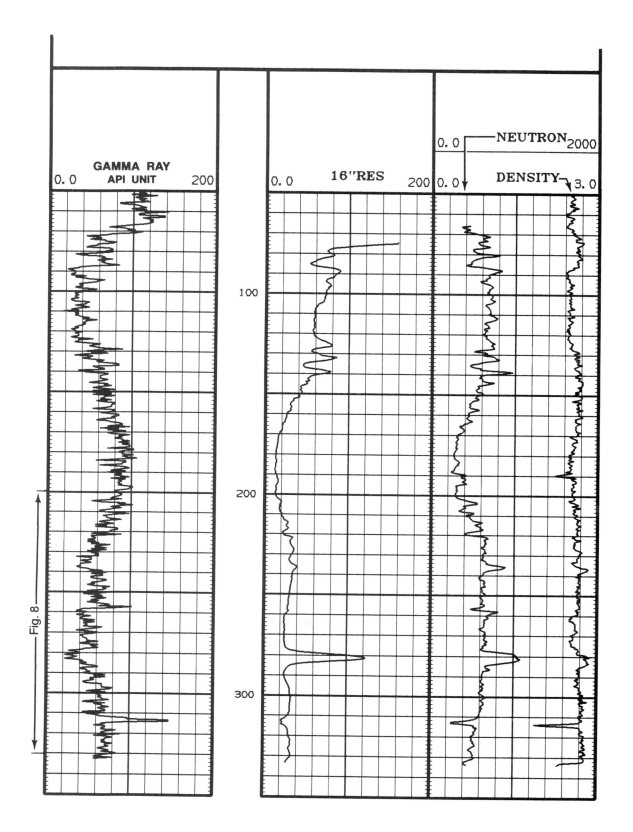

Figure 9. Log response of ARCO Well 82-8. Interval represented in Figure 8 is shown. Note lack of resistivity response in siltstone (except for sharp resistivity increase due to a septarian nodule at 280 ft).

primary structures (such as cross bedding), and is only moderately burrowed.

The Ferron shelf clastic interval is distinguished by its thickness (> 100 ft; 30 m) and its lack of sedimentary structures. Characteristics of the siltstone portions are starved ripples with occasional small (1-10 cm) interbeds of coarse silt or very fine sandstone indicating storm deposition. Other characteristic sedimentary structures are graded beds (see photographs of boxes 41 through 43 and box 56). These beds were probably deposited during the waning stages of storms (as indicated by partial hummocky sequences). Common trace fossils (Table 1) are toward the top. Overlying the bioturbated siltstone is a laminated and rippled mudstone which is characteristic of suspension sedimentation.

The gamma-ray log response (Fig. 11) decreases slightly upward at the base and increases slightly toward the top of the sequence as the amount of shale varies. Bentonite beds are common in this facies as are calcite-cemented septarian nodules. Spikes in the gamma-ray pattern at 258 ft and 312 ft mark bentonite layers, and the resistivity spike at 280 ft records a septarian nodule observed in core. This interval is interpreted as the distal part of a shelf sand body which was reworked from the Vernal Delta complex located about 50 mi north of the cored sequence. Outcrop observations by one of us (Ossian) suggest that this bioturbated sandstone/siltstone unit coarsens to the north. In addition, sand waves were observed in outcrop north of the core sites and at the lower Ferron bench horizon. These sand waves indicate a general south-southwest transport direction (see Fig. 6A).

Name	Description	Environment
Asterosoma	Star-shaped structures having elevated, often asymmetrical, centers. In cross section, concentric laminae of sand and clay packed about and below a central tube. Feeding and resting trace of decapod crustaceans.	Low-energy, shallow water. Tidal flat, lagoon, shoreface to offshore.
Chondrites	Branching burrow system, mostly horizontal or inclined. Small, constant-diameter, smooth-walled, mucous-lined burrows. Worm feeding trace.	Still water, variable depth. Nearshore to basin.
Helminthoida	Tightly looping fecal ribbons of a grazing worm.	Shallow water, low energy. Nearshore to transition zone; also deep sea.
Ophiomorpha	Cylindrical pipes (1-5 cm diameter) having smooth interior and pelleted exterior. May be branching, horizontal or vertical. Dwelling of mud shrimp or other crustacean.	Shallow, high-energy water in littoral or sublittoral zone on beaches, shoals, intertidal flats. Brackish to marine in sandy substrate.
Planolites	Sand-filled burrows without distinct internal structure. Worm grazing trace.	Tidal flat, nearshore, and transition zone.
Rhizocorallium	Horizontal or inclined spreite-filled U-shaped burrow. Grazing and feeding trace.	Lagoon and backshore; nearshore to transition zone.
Teichichnus	Vertical tabular structures made of stacked biogenic laminae.	Lagoon and backshore; shoreface to deep basin.
Terebellina	Unsculpted, grain-lined tubes ("donut" burrows), usually occurring in clusters.	Shoreface, nearshore, transition zone.
Thalassinoides	Large branching burrow and tunnel system similar to Ophiomorpha but lacking pelleted exterior. Dwelling place of crustaceans.	High-energy, just below intertidal zone. Lagoon; shoreface to transition zone.

FERRON PROJECT ARCO WELL 82-8 210-214 FT. — BOX 35 — TOP

FERRON PROJECT ARCO WELL 82-8 215-220 FT. — BOX 36

FERRON PROJECT ARCO WELL 82-8 222-227 FT. — BOX 37

BOTTOM

Boxes 35-37. Bioturbated silty sandstone in sharp contact with interbedded siltstone and mudstone, indicative of a back plume environment on the shelf.

345

FERRON PROJECT
ARCO WELL 82-8
228-231 FT.

BOX 38

FERRON PROJECT
ARCO WELL 82-8
240-243 FT.

BOX 41

FERRON PROJECT
ARCO WELL 82-8
244-246 FT.

BOX 42

TOP

BOTTOM

Boxes 38, 41-42. Bioturbated silty sandstone consisting of upward fining siltstone sequences and graded beds which are indications of storm deposits.

346

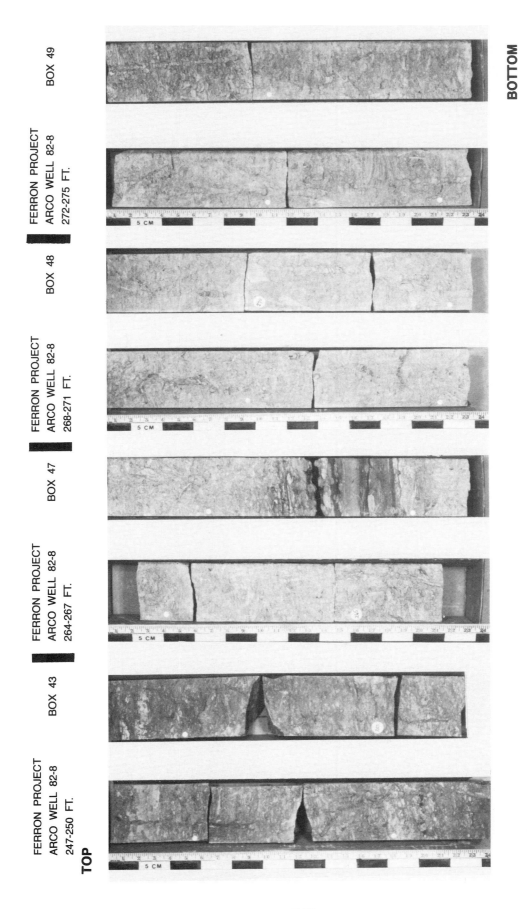

FERRON PROJECT
ARCO WELL 82-8
247-250 FT.

BOX 43

FERRON PROJECT
ARCO WELL 82-8
264-267 FT.

BOX 47

FERRON PROJECT
ARCO WELL 82-8
268-271 FT.

BOX 48

FERRON PROJECT
ARCO WELL 82-8
272-275 FT.

BOX 49

TOP

BOTTOM

Boxes 43, 47-49. Bioturbated sandy siltstone to silty sandstone. Teichichnus is dominant trace fossil in this group of cores: note side view of Teichichnus in left core of Box 49. Graded beds are indicated in Box 47.

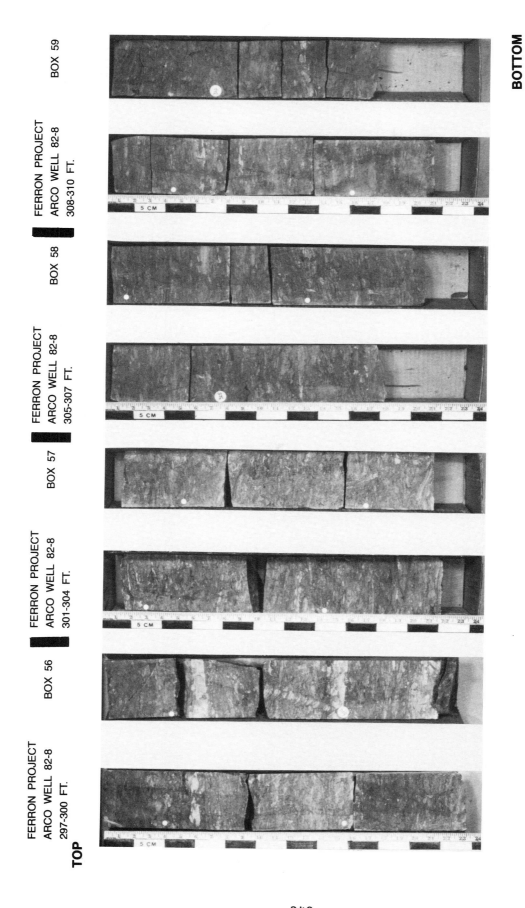

TOP

FERRON PROJECT
ARCO WELL 82-8
297-300 FT.

BOX 56

FERRON PROJECT
ARCO WELL 82-8
301-304 FT.

BOX 57

FERRON PROJECT
ARCO WELL 82-8
305-307 FT.

BOX 58

FERRON PROJECT
ARCO WELL 82-8
308-310 FT.

BOX 59

BOTTOM

Boxes 56-59. Bioturbated sandy siltstone to silty sandstone. The transition is gradual. Note thin sandstone interbeds, probably indicative of storm deposit on the shelf. Also note Teichichnus in Box 56.

348

Boxes 62-64. Cross-bedded fine sandstone in Boxes 63 and 64 indicating upper shoreface deposits just seaward of the surf zone. The top half of Box 63 and lower half of Box 62 consist of burrowed muddy sandstone, indicative of reworking during marine transgression. The top part of Box 62 is a laminated mudstone deposited during a local transgression.

349

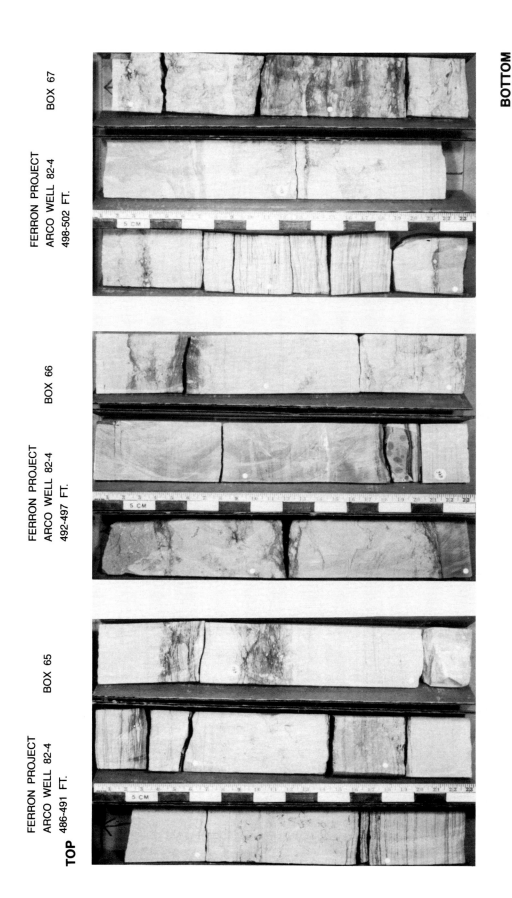

TOP

BOTTOM

FERRON PROJECT
ARCO WELL 82-4
486-491 FT.

BOX 65

FERRON PROJECT
ARCO WELL 82-4
492-497 FT.

BOX 66

FERRON PROJECT
ARCO WELL 82-4
498-502 FT.

BOX 67

Boxes 65-67. Horizontally laminated very fine sandstone. Gradational contact at base as sand content increases. These are probably storm beds in a lower shoreface setting. Note sharp contact with overlying cross-bedded fine sandstone at top of Box 65.

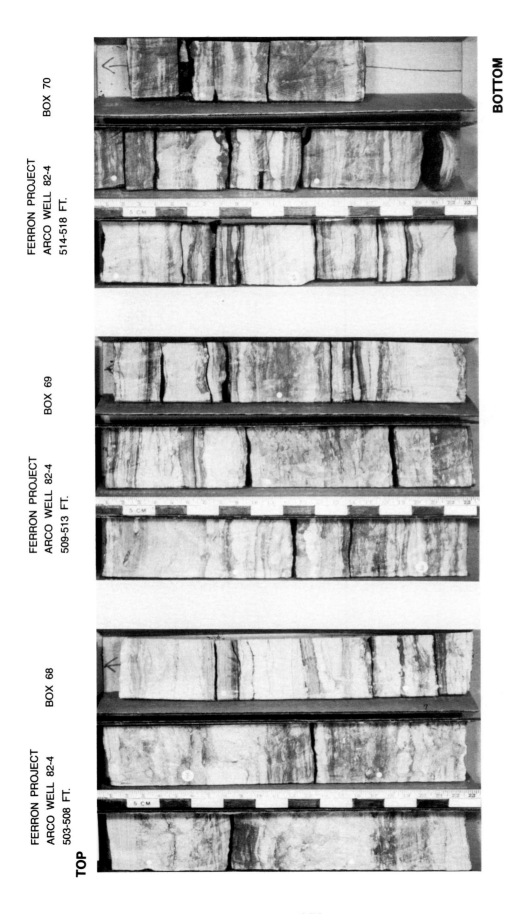

FERRON PROJECT
ARCO WELL 82-4
514-518 FT.

BOX 70

FERRON PROJECT
ARCO WELL 82-4
509-513 FT.

BOX 69

FERRON PROJECT
ARCO WELL 82-4
503-508 FT.

BOX 68

TOP

Boxes 68-70. Interbedded sandstone and siltstone in gradational contact with rippled siltstone (Box 70) and increasing upward in sand content. Represents a transition between offshore and lower shoreface deposits. The fine sandstone beds were probably deposited during storms.

FERRON PROJECT
ARCO WELL 82-4
519-524 FT.

BOX 71

FERRON PROJECT
ARCO WELL 82-4
525-530 FT.

BOX 72

FERRON PROJECT
ARCO WELL 82-4
531-535 FT.

BOX 73

TOP

BOTTOM

Boxes 71-73. Rippled siltstone consisting of interbedded mudstone and sandy siltstone with moderate burrowing. Thicker laminated sandstone beds may indicate welded plume deposits.

ARCO Oil and Gas Company Core 82-4

ARCO 82-4 (Figs. 10, 11) is from a more proximal part of the delta complex. The entire core represents a progradational facies tract. Shelf sediments at the base are overlain by delta front, lower delta plain, and upper delta plain deposits. ARCO 82-4 is located in a more proximal portion of the delta complex than was ARCO 82-8. The interval represented in Figure 8 is an upward-coarsening sandstone facies assemblage from the delta front shoreface. Component lithofacies (from the base) are (1) laminated mudstone, (2) rippled siltstone, (3) interbedded sandstone and siltstone, (4) horizontally-laminated very fine sandstone, and (5) cross-bedded, fine sandstone.

The rippled siltstone interval gradationally overlies laminated mudstone. It consists of thin beds of siltstone and very fine sandstone, and is about 20 ft (6.3 m) thick in the ARCO 82-4 core. The most common recognizable trace fossil in this extensively burrowed lithofacies is Teichichnus. Core features indicate that this location was subject to a relatively low-energy regime during fair weather conditions. Interbeds of sandstone at the base of the sequence may represent welded sandy shelf deposits overlain by the siltier eddy deposits (Fig. 3).

The interbedded sandstone and siltstone lithofacies consists of 4-6 in. (10-15 cm), upward-fining horizontally-laminated sandstones interbedded with rippled and burrowed siltstones. The siltstone beds are very thin (< 3 in; 75 cm) and overlie thicker (as much as 1 ft; 0.3 m) beds of horizontally-laminated, very fine sandstone. Upper surfaces of sandstone beds are either rippled or burrowed. The entire

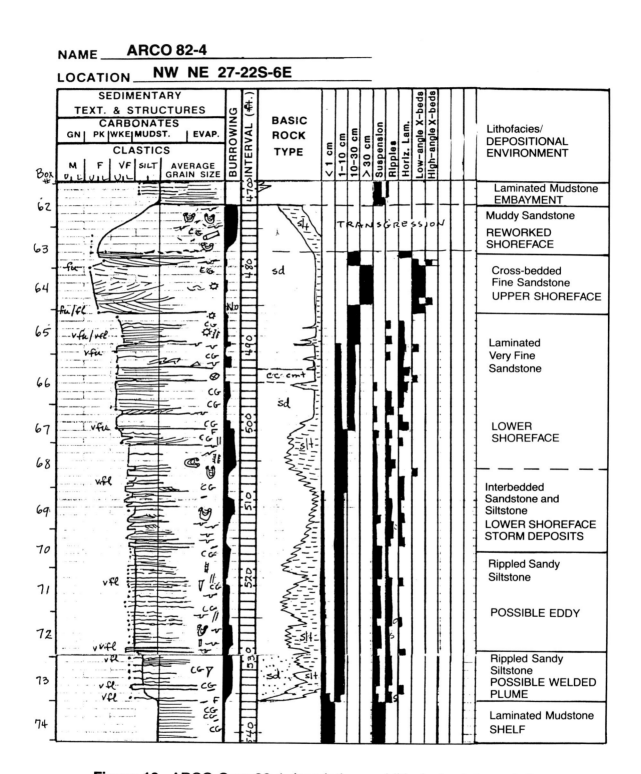

Figure 10. ARCO Core 82-4 description and lithofacies interpretation.

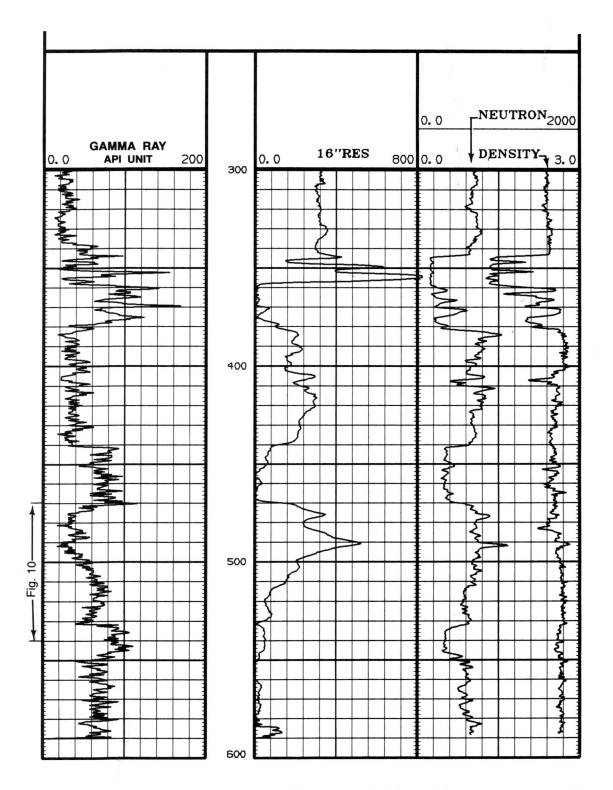

Figure 11. Log response of a segment from ARCO Well 82-4. Interval represented in Figure 10 is shown. Resistivity increases significantly in porous sandstone of the upper shoreface.

interbedded sandstone and siltstone sequence is less than 10 ft (3.2 m) thick in ARCO 82-4 (see photographs).

This lithofacies closely resembles the hummocky cross-stratified sequence described by Dott and Bourgeois (1982). A complete hummocky sequence consists of a scoured base overlain by low-angle to flat beds and ripples with burrows and a mud drape at the top. Even though the characteristic large-scale antiform hummocks probably cannot be distinguished in core, their presence in outcrop at this same stratigraphic position indicates lower shoreface deposition below fair-weather wave base.

The horizontally-laminated very fine sandstone is about 20 ft (6.4 m) thick in ARCO 82-4 (Fig. 8). Some beds are separated by burrowed zones or ripples, and the laminae are sometimes faint. These beds might be the amalgamated hummocky sequences of Dott and Bourgeois (1982). Ophiomorpha burrows are common near the top, and macerated plant material occurs throughout. Some beds have been syndepositionally deformed, suggesting that episodic depositional events (such as storms) caused rapid influx of sediments and subsequent dewatering or slumping of water-saturated sand.

A thin (< 10 ft; 3.2 m), cross-bedded fine-grained sandstone caps this facies assemblage in ARCO 82-4. Cross-bedding was formed by migrating bedforms within the surf zone. The surface of this sandstone interval is burrowed. The most common trace fossils throughout are *Thalassinoides* and Teichichnus burrows at the very top of the interval. Ophiomorpha burrows are sparse, but macerated plant material is abundant, particularly near the top.

The wire-line log clearly reflects a coarsening-upward progradational sequence (Fig. 9). The gamma-ray log is characterized by a general upward decrease in count rate on top of a fairly sharp base and an initial decrease in count rate. The serrate pattern reflects silty and shaly interbeds. Resistivity increases upward in the section as the amount of shale and silt decreases.

This sequence is interpreted as a progradational shoreface deposit overlying shelf deposits, with a sequence of possible "welded" shelf sand deposits at its base. With the exception of basal thin sandstone interbeds, the interval resembles other progradational shoreface sequences observed both in modern environments (Morton and McGowen, 1980) and in the Cretaceous Western Interior Seaway (Boyles, 1983, Noe, 1984; Cant, 1984).

Conclusions and Economic Potential

1. Core 82-8 is a sequence from the distal part of a shelf sand "plume" derived from the northern Vernal Delta. Sediments are very fine-grained, bioturbated, and lack sedimentary structures. The interval is about 80 ft (28 m) thick.

2. Core 82-4 consists of the delta front, shoreface, and welded shelf "plume" deposits originating from a younger southern delta. Grain size coarsens upward from sandy silt to fine sandstone.

3. The bioturbated siltstone interval (seen in the cores) underlying the prograding deltaic deposits provide a significant proximity indicator for thicker, more porous sandstones of the main shelf sand body sediments (Patterson, 1983; Palmer and Scott, 1984).

4. Shelf sand body sediments interfinger with hydrocarbon source rocks and in the Ferron example are overlain and sealed by lower porosity siltstones of the lower shoreface and prodelta sediments.

5. Shelf "plume" sequences have not been generally recognized in exploration efforts. Recent work suggests that these settings may provide an important new direction in the search for stratigraphic traps in areas previously thought to be noneconomic.

References

Boyles, J. M., 1983, Depositional history and sedimentology of Upper Cretaceous Mancos Shale and Lower Mesaverde Group, northwestern Colorado--migrating shelf-bar and wave dominated shoreline deposits (Ph.D. dissert: Austin, Univ. Texas, Ph.D. dissertation, 270 p.

Boyles, J. M., and Scott, A. J. 1982, A model for migrating shelf-bar sandstones in upper Mancos Shale (Campanian), northwestern Colorado: Amer. Assoc. Petrol. Geol. Bull. v. 66, no. 5, p. 491-508.

Cant, D. J., 1984, Development of shoreline-shelf sand bodies in a Cretaceous epeiric sea deposit: Jour. Sed. Petrol., v. 54, p. 541-56.

Chamberlain, C. K., 1978, Recognition of trace fossils in cores, in Bassan, P.B., ed., Trace fossil concepts, SEPM Short Course No. 5, p. 119-66.

Coleman, J. M., Roberts, H. H., Murray, S. P., and Salama, M., 1981, Morphology and dynamic sedimentology of the eastern Nile Delta shelf: Mar. Geol. v. 42, p. 301-26.

Cotter, E., 1975a, Deltaic deposits in the Upper Cretaceous Ferron Sandstone of Utah, in Broussard, M.L., ed., Deltas, models for exploration, Houston Geol. Soc., p. 471-84.

Cotter, E., 1975b, Late Cretaceous sedimentation in a low-energy coastal zone, the Ferron Sandstone of Utah: Jour. Sed. Petrol., v. 45, p. 669-85.

Dott, R. H., and Bourgeois, J., 1982, Hummocky stratification, significance of its variable bedding sequences: Geol. Soc. Amer. Bull., v. 93, p. 663-80.

Ekdale, A. A., Bromley, R. G., and Pemberton, S. G., 1984, Ichnology, the use of trace fossils in sedimentology and stratigraphy: SEPM Short Course No. 15, 317 p.

Galloway, W. E., 1975, Process framework for describing the morphologic and stratigraphic evolution of deltaic depositional systems, in Broussard, M. L., Deltas, models for exploration): Houston Geol. Soc., p. 87-98.

Morton, R. A., and McGowen, J. H., 1980, Modern depositional environments of the Texas coast: Austin, Univ. Texas, Bur. Econ. Geol. Guidebook 20, 167 p.

Nittrouer, C. A., 1981, Sedimentary dynamics on continental shelves: Elsevier Publ. Co.

Noe, D. C., 1984, Variations in shoreline sandstones from a Late Cretaceous interdeltaic embayment, Sego Sandstone (Campanian), northwestern Colorado: Austin, Univ. Texas, Master's thesis, 127 p.

Palmer, J. J., and Scott, A. J., 1984, Stacked shoreline and shelf sandstone of LaVentana Tongue (Campanian), northwestern New Mexico: Amer. Assoc. Petrol. Geol. Bull., v. 68, p. 74-91.

Patterson, J. E., 1983, Exploration potential and variations in shelf plume sandstones, Navarro Group (Maetrichtian), east central Texas: Austin, Univ. Texas, Master's thesis, 91 p.

Reading, H. G., ed., 1978, Sedimentary environments and facies: Elsevier, 569 p.

Ryer, T. A., and McPhillips, M., 1983, Early Late Cretaceous paleogeography of east-central Utah, in Reynolds, M. W., and Dolly, E. D., eds., Mesozoic paleogeography of the west central United States: Rocky Mt. Sec., SEPM Symposium 2, p. 253-72.

Stubblefield, W. L., McGrain, D. W., and Kersey, D. G., 1984, Recognition of transgressive and post-transgressive sand ridges on the New Jersey continental shelf, in Tillman, R. W., and Siemers, C. T., eds., Siliciclastic shelf sediments: Soc. Econ. Pal. Min. Spec. Publ. 34, p. 1-24.

Swift, D. J., McKinney, T. F., and Stahl, L., 1984, Recognition of transgressive and post-transgressive sand ridges on the New Jersey continental shelf--discussion, in Tillman, R. W., and Siemers, C. T., eds., Siliciclastic shelf sediments: Soc. Econ. Pal. Min. Spec. Publ. 34, p 25-36.

Thompson, S. L., 1985, Ferron Sandstone Member of the Mancos Shale, Turonian mixed energy deltaic system: Austin, Univ. Texas, Master's thesis, 165 p.

Thompson, S. L., and Ossian, C. R., 1985, Ferron Sandstone Member of Mancos Shale, a Turonian mixed energy deltaic system (Abstract): SEPM Annual Midyear Mtg. Abstracts, v. 2, p. 89.

SEDIMENTARY FACIES OF THE UPPERMOST WILCOX SHELF-MARGIN TREND: SOUTH-CENTRAL LOUISIANA

Philip Lowry, Rowdy C. Lemoine, and Thomas F. Moslow

Basin Research Institute and Louisiana Geological Survey, Louisiana State University, Baton Rouge, Louisiana 70803

Abstract

Fordoche Field, which has estimated reserves in excess of 90 million barrels of oil and gas, contains several stacked sandstones that are part of a paleo shelf-margin trend within the downdip uppermost Wilcox of south-central Louisiana. Commonly referred to as the "Deep Wilcox," this trend contains at least one submarine canyon-fill and is coincident with the underlying Cretaceous carbonate reef trend. Thus, antecedent topography has significantly influenced the patterns of sedimentation and preserved sand-body geometry within this downdip Wilcox trend.

The main reservoir intervals appear "blocky" on electric logs, average 30 to 40 ft (9.14-12.2 m) thick, and are laterally continuous in an east-west (strike) direction over a distance of 40 mi (64.4 km) and at least 6 mi (8 km) in a north-south (dip) direction. Analysis of over 300 ft (91.4 m) of conventional core from the W8 Sandstone within Fordoche Field suggests deposition in a wave-dominated shoreface environment at or near the shelf margin. Use of the term "shelf-margin delta" or "shelf-edge delta" has been avoided only because there is no direct evidence in the cores of a developed fluvial system; however, the sandstone bodies are believed to be somewhat analagous to late Quaternary Gulf Coast shelf-margin deltas.

Six major lithofacies ("A" through "F") are identified within the W8 Sandstone. Facies A through C, which constitute over 90% of the cored sequence, represent the initial progradation of inner-shelf and shoreface sandstones over outer-shelf and upper-slope mudstones. Facies D and E are sandstones and sandy mudstones that represent the remnants of a transgressive event that reworked upper-shoreface and foreshore deposits. Facies F is composed of mudstones that represent suspension sedimentation in an outer-shelf environment associated with the ongoing transgression.

Optimum reservoir quality is associated with the less bioturbated upper portion of facies C. Greatest permeability and porosity values occur, however, in thin, discrete sandstone beds (5-30 cm thick) (interpreted as tempestites) in facies B underlying the main reservoir sandstone. Although several of these beds produce hydrocarbons, most are not laterally continuous and are frequently bioturbated, which impairs reservoir homogeneity.

Introduction

Wilcox in Louisiana

The Wilcox Group (Paleocene-Eocene) (Fig. 1) is a well-established and prolific hydrocarbon-producing sequence in the northern Gulf of Mexico basin. Locally, the Wilcox is over 4,000 ft (1,220 m) thick and represents the initial influx of terrigeneous clastic sediments to the Louisiana region of the basin during the early Tertiary. The majority of Wilcox oil and gas production within Louisiana originates from relatively shallow (3,000-6,000 ft; 914-1,828 m), dip-oriented sandstone reservoirs in the northern region of the state (Craft, 1966). Toward the south and downdip from this developed region is a strike-oriented trend extending 140 mi (225 km) from Livingston Parish in the east, westward to the Texas-Louisiana state line (Fig. 2). The average depth to the producing intervals is 10,000-14,000 ft (3,048-4,267 m). Included in this downdip "Deep Wilcox" trend are the initial discovery at Fordoche Field and the important recent discoveries at Lockhart Crossing and Livingston fields (Self et al., 1985).

The purpose of this study is to examine the nature, distribution, and reservoir quality of sedimentary facies from a producing interval within this downdip trend in south-central Louisiana on the basis of sedimentologic analysis of conventional cores from Fordoche Field. In addition to core analysis, well-log correlations are used to construct cross sections that provide a regional stratigraphic perspective of the producing interval.

Stratigraphic Nomenclature

Owing to a limited number of surface exposures and pervasive lateral facies changes, correlation of formations from surface to

ERA	SYSTEM		SERIES	ROCK UNIT
CENOZOIC	TERTIARY	QUAT	HOLOCENE	RECENT
			PLEISTOCENE	
		NEOGENE	PLIOCENE	CITRONELLE FM.
			MIOCENE	FLEMING FM.
				CATAHOULA FM.
		PALEOGENE	OLIGOCENE	VICKSBURG GRP.
			EOCENE	JACKSON GRP.
				CLAIBORNE GRP.
				WILCOX GRP.
			PALEOCENE	MIDWAY GRP.

MLE, Carto. Sect., LSU

Figure 1. Generalized Gulf Coast Cenozoic stratigraphic column.

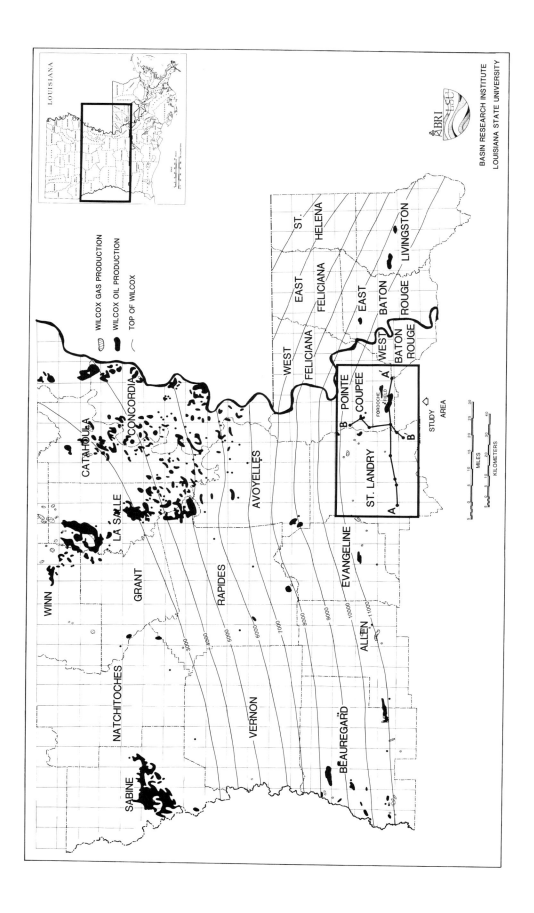

Figure 2. Location map of oil and gas fields in central Louisiana showing structural contours on top of the Wilcox. The enclosed region in the lower part of the map is sthe study area for this investigation and shows the location of regional strike and dip cross sections in Figures 5 and 6 (modified from Oil and Gas Map of Louisiana, Louisiana Geological Survey, 1981).

subsurface has met with minimal success in Louisiana. Consequently, many subsurface geologists have adopted a simplistic threefold division of the Wilcox section based on electric-log pattern recognition (Albach, 1979; Coates, 1979; Mulcahy, 1981; Rogers, 1983). The lower, middle, and upper Wilcox are interpreted as fluvial-dominated deltaic, deep-marine, and wave-dominated deltaic systems, respectively (Fisher and McGowen, 1967; Galloway, 1968). In practice, recognition of this threefold division within Louisiana may obscure true stratigraphic relationships. In this paper, reference is made to locations within the section as being lowermost or uppermost simply to indicate relative position in the section and thus avoid possible unwarranted genetic connotations.

The most basinward trend of Wilcox production is frequently termed the "Deep Wilcox" because of the subsea depth of production intervals (Berg and Tedford, 1977; Edwards, 1980). Since most hydrocarbon production from this trend originates from the uppermost Wilcox sandstones, preference is given here to the term "downdip uppermost Wilcox" to avoid confusion regarding relative depth of reservoirs within the section.

Regional Geologic Setting

The Wilcox Group is the lowermost portion of a thick sequence of Tertiary clastic sediments deposited along the northern rim of the Gulf of Mexico basin. A narrow band of Wilcox outcrops extends from Alabama to Texas, but for the most part, the majority of the section is confined to the subsurface. Consequently, much of the present understanding of Wilcox depositional history is based primarily on well-log correlations and subsurface facies analysis (Fisher and McGowen, 1967; Galloway, 1968).

Detailed regional paleogeographic and stratigraphic relationships of the uppermost Wilcox remain unclear. However, three important aspects of the regional geologic setting of the present study area may help improve our understanding of the paleogeography and stratigraphy of the uppermost Wilcox. First, Anisgard (1970) determined through the analysis of foraminiferal assemblages that much of the downdip Wilcox trend was deposited in inner- to middle-neritic marine conditions where average water depths were 100 ft (30.5 m). The foraminiferal assemblages were interpreted as characteristic of turbid, poorly oxygenated marine waters.

Secondly, Winker and Edwards (1983) documented the effect of a previously established Cretaceous shelf margin on all subsequent deposition. From early Cretaceous to the Paleocene, shelf margins remained within a relatively confined zone (Fig. 3) (Hendricks and Wilson, 1967; Stehli et al., 1972; Christina and Martin, 1979; Winker, 1982). The early Cretaceous carbonate shelf-edge reef trend formed a stable, well-defined margin. When this margin was eventually overlain by clastic sediments, the flexure controlled the location of successive shelf edges and created a regional zone of instability.

The third aspect of the regional geologic setting pertinent to this study area is the discovery of a large mudstone-filled channel in St. Landry Parish (Fig. 3) (Lowry et al., in press; McCulloh and Eversull, in press). This feature is interpreted to represent a Wilcox submarine canyon system and appears similar to other thick mudstone channel fills in Texas (Hoyt, 1959; Chuber and Begeman, 1982). The "St. Landry"

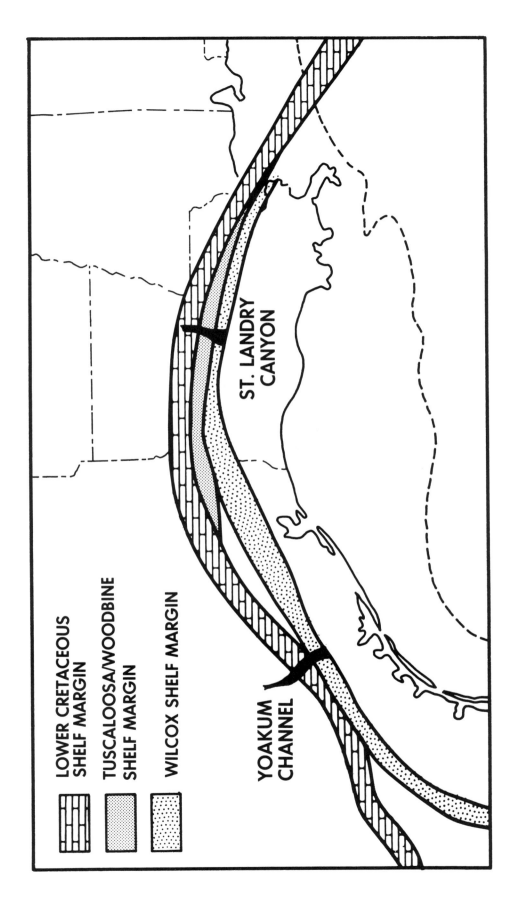

Figure 3. Map of lower Cretaceous, Tuscaloosa/Woodbine, and Wilcox shelf-margin trends in the norther Gulf Coast with locations of Yoakum Channel and St. Landry Canyon (modified from Winker, 1982).

LOWER CRETACEOUS SHELF MARGIN

TUSCALOOSA/WOODBINE SHELF MARGIN

WILCOX SHELF MARGIN

YOAKUM CHANNEL

ST. LANDRY CANYON

channel and the W8 Sandstone are stratigraphic equivalents. Given the modern submarine canyons (e.g., Mississippi Canyon) from at or near the shelf margin, data from this study, when merged with previous paleontological and stratigraphic findings, suggest that sediments of the downdip Wilcox trend in Louisiana were deposited close to the shelf margin.

Fordoche Field

Development History

Wilcox production in Fordoche Field was first established in November 1965 from the Sun Kent #1 discovery well (Pierson, 1970). The Kent #1 well is significant in that it helped establish a new production trend in the downdip Wilcox in south-central Louisiana. Initial production was 411 barrels per day from two intervals, one at 13,784-13,796 ft (4,201-4,205 m) and the other at 13,933-13,983 ft (4,247-4,262 m). Fordoche Field occurs within a "deep-seated" anticline associated with a major growth fault. There are five producing intervals within the field, referred to as the W4, W5, W8, W12, and W15 sandstones (Fig. 4). This paper examines the sedimentary characteristics and reservoir quality of the W8 Sandstone.

Estimated reserves-in-place for Fordoche Field are approximately 91 million barrels (91 mm bbl). However, as of 1983, ultimate recovery after reservoir stimulation was estimated at only 27.6 million barrels (27.6 mm bbl) or 30%. Sun Oil Company, the major operator in the field, began a miscible gas enhanced-recovery project that has proved very successful. Nitrogen injection from three wells greatly increased reservoir pressures and flow rates, especially for the W8 Sandstone. The W8 interval has yielded 11.7 million barrels (11.7 mm bbl) of highly

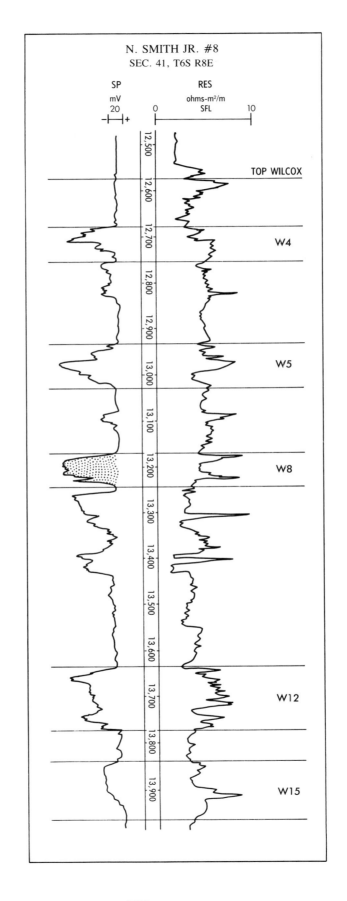

Figure 4. Wilcox "type-log" for Fordoche Field from the N. Smith Jr. #8 well. Cored sequences from the W8 Sandstone (stippled) are examined in this study. Main producing intervals are labeled (modified from Eckles et al., 1981).

volatile oil (45.8° API gravity), which is approximately 31% of the estimated oil in place. Information gained from this study pertaining to the input of depositional controls on reservoir quality and performance should be helpful in planning any further enhanced-recovery projects.

Reservoir Geometry and Growth Faulting

The W8 Sandstone is one of three thick sandstone intervals that occur within a sedimentary package 40 mi long and almost 300 ft (91.4 m) thick (Fig. 5). All three sandstones rapidly pinch out over 6-7 mi (9-11 km) in a basinward direction (Fig. 6). Updip, the character of the package, as determined from well logs, is significantly different and consists of interbedded sandstones and shales (Fig. 6).

It is clearly evident from Figure 6 that faulting has had a significant influence on deposition in this area. Both of the faults depicted on the dip cross section are major syndepositional faults (growth faults), which are a common occurrence in Gulf Coast shelf margins (Lehner, 1969; Busch, 1975; Jackson and Galloway, 1984). Fault A (Fig. 6) is interpreted to have been active during Wilcox deposition, whereas growth on fault B occurred after Wilcox deposition. Numerous mechanisms have been proposed for the initiation of growth faulting, many of which occur at the shelf edge; however, their net result is a thickening of section on the downthrown side and the development of rollover anticlines (Durham and Peoples, 1956; Martin, 1977). Such rollover anticlines are perhaps the most common exploration prospect in the Gulf Coast.

Although the influence of growth faulting can be demonstrated on a regional scale (Thorsen, 1963), the control on a specific depositional event, which may only be represented by an interval 10-15 m thick, is

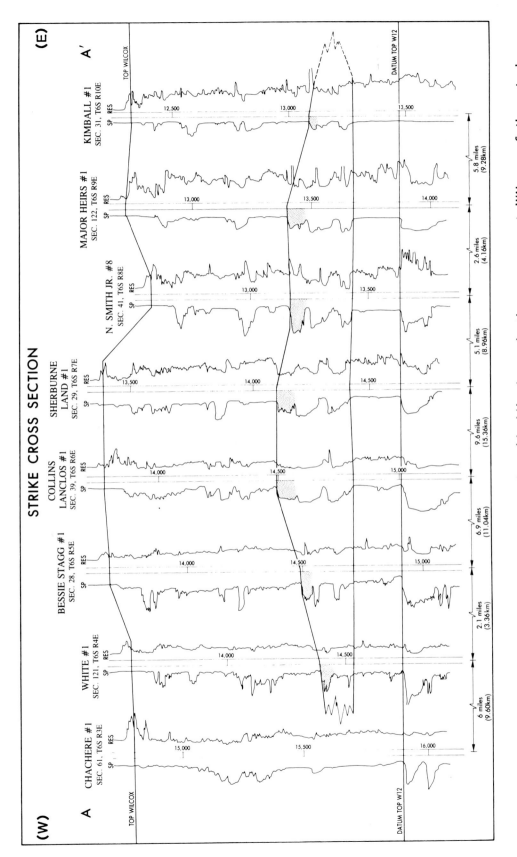

Figure 5.   Regional strike-oriented cross section (A–A') through the uppermost Wilcox of the study area (see Fig. 2 for location of section). The W8 Sandstone is indicated by a stippled pattern. Note that the W8 overlies a thick sandstone sequence that pinches out along strike to the east and west.

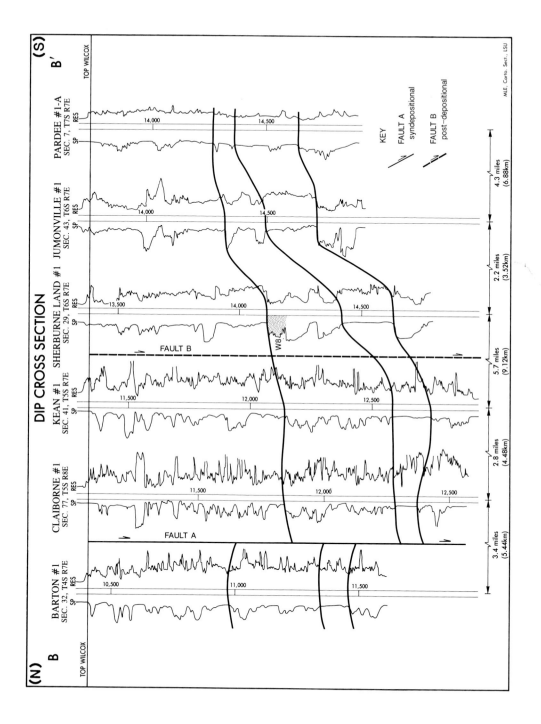

Figure 6. Regional dip-oriented cross section (B-B') through the uppermost Wilcox of the study area (see Fig. 2 for location of section). The W8 Sandstone is stippled in the Sherburne Land #1 well, which was used to define the along-strike continuity of the W8 in cross section A-A' (Fig. 5). Note rapid lateral facies changes in the W8 in a downdip and updip direction.

less clear. Edwards (1981) suggests that growth faulting played an important role in the development of upper Wilcox shelf-edge deltas in southern Texas, whereas Suter and Berryhill (1985) found it to have relatively little structural influence on the development of Gulf Coast late Quaternary shelf-margin deltas. This study shows that although growth faults are apparently important, they may not be a necessary precursor in the development of strike-oriented sand bodies at the shelf edge.

Cored Sedimentary Facies

Vertical Sequence

We used approximately 300 ft of conventional cores from five different wells to examine the W8 sandstone of Fordoche Field (Fig. 7). A detailed sedimentologic description was written for all cores in order to determine the following: (1) the sedimentary characteristics and recognition criteria of distinct lithofacies, (2) the processes responsible for deposition of the sediments, (3) the environment in which the sediments were deposited, and (4) the relationship between primary patterns of sedimentation and reservior quality.

Six distinct sedimentary facies (A-F) are recognized in the W8 Sandstone interval. On the basis of core analysis, we have compiled a summary of sedimentary characteristics (lithology and physical and biogenic sedimentary structures), sequence characteristics, and relative reservoir quality for all facies (Table 1) and describe them in detail below. A composite sedimentary sequence showing the vertical succession of facies in the W8 Sandstone has been compiled principally from analysis of core from the N. Smith Jr. #8 well near the central portion of Fordoche Field (Figs. 7, 8).

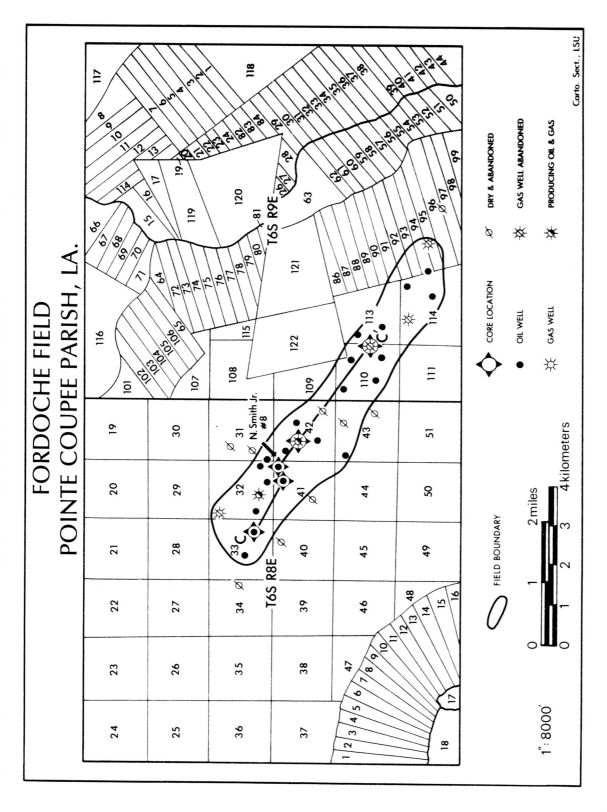

Figure 7. Map of Fordoche Field. Note locations of cored wells including the N. Smith Jr. #8 well in T6S, R8E, sec. 41. Cross section C-C' is through the central axis of the field.

TABLE 1. SEDIMENTARY CHARACTERISTICS AND RELATIVE RESERVOIR QUALITY OF CORED FACIES

	Facies A	Facies B	Facies C	Facies D	Facies E	Facies F
LITHOLOGY	Mudstone (50% clay, 45% silt) thin(1-10mm) silt lenses dispersed throughout interval, glauconite <1%; localized concentrations of pyrite <2%.	Muddy, very fine grained sandstone (mean 90-100μ, max 125μ), with thin(5-30cm)silty, very-fine grained sandstone interbeds; glauconite 2-3%; pyrite <1%.	Silty, very fine grained, well sorted sandstone (mean 100-125μ, max 150μ); glauconite <2%.	Very fine grained, calcareous sandstone (mean 100-125μ, max 150μ); diverse small (5-10mm) carbonate clasts.	Very fine sandy mudstone (40% sand); glauconite 10-15%; siderite concretions.	Mudstone (40% clay, 55% silt); glauconite 1-3%, localized concentrations of pyrite <2%.
PHYSICAL SEDIMENTARY STRUCTURES	Lenticular bedding soft sediment deformation; rare load-casted ripples.	Horizontal to sub-horizontal laminations; rare low angle truncation surfaces.	Massive appearing (burrowed?); rare horizontal laminations.	Massive appearing.	Rare horizontal laminations.	Horizontal laminations.
BIOGENIC SEDIMENTARY STRUCTURES	Rare burrowing; low diversity of burrow types(Chondrites, Terebellina).	Common burrowing; moderate diversity of burrow types (Teilchichnus, Terebellina, Planolites, Chondrites).	Abundant burrowing; low diversity of burrow types (Ophiomorpha).	No distinct burrow types.	Bioturbated (>75% burrowed).	Rare burrowing; low diversity of burrow types (Terebellina).
SEQUENCE CHARACTERISTICS	Lower contact missing, gradational upper contact; decrease in soft-sediment deformation and increase in burrowing upwards.	Sandstone beds exhibit sharp lower contacts and burrowed upper contacts; increase in thickness and frequency of occurrence of discrete sandstone beds upwards.	Lower contact gradational, sharp upper contact; increase in sand content and decrease in degree of burrowing upwards.	Lower and upper contacts sharp.	Lower contact sharp, gradational upper contact; decrease in glauconite and increase in mud contents upwards.	Lower contact gradational, upper contact sharp; increase in silt content upwards.
RESERVOIR QUALITY *	Not a reservoir unit.	Discrete sandstone beds $\emptyset$.....10-23% K.....10-98md S_o.....15-23% Muddy sandstone beds $\emptyset$.....10-15% K.....0.1-3md S_o.....5%	$\emptyset$.....15-22% K.....0.5-14md S_o.....20-30% (Best reservoir quality facies)	$\emptyset$.....4% K......0.1md S_o.....7% (possible diagenetic seal)	$\emptyset$.....6-12% K.....0.1md S_o.....0-4%	Not a reservoir unit.

*Based on Porosity ($\emptyset$), Permeability (K), and Oil Saturation (S_o).

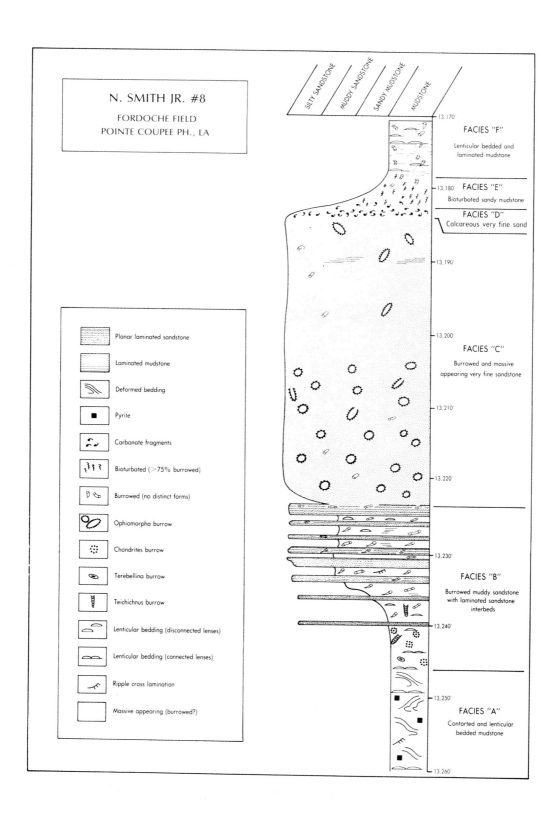

Figure 8. Schematic core description for N. Smith Jr. #8 well. This sequence is interpreted to be characteristic of the W8 Sandstone.

The vertical sequence of facies in the W8 interval is dominated by a relatively thick (35-40 ft) burrowed and massive-appearing sandstone (facies C) that shows an upward increase in sand percentage. Underlying and gradational with facies C is a burrowed muddy sandstone with laminated sandstone interbeds (facies B). Facies B and C are the major reservoir units in the W8 interval. Facies C is capped in almost all producing wells by a thin (2-3 ft), calcareous, tightly cemented, very fine-grained sandstone that appears to be a potentially important diagenetic reservoir seal. Underlying the sandstone units in the W8 interval are contorted and lenticular-bedded mudstones of facies A. Bioturbated sandy mudstones (facies E) and lenticular-bedded to laminated mudstones (facies F) overlie the sandstone units (Fig. 8).

A complete succession of photographs showing sedimentary structures and lithologies for the entire cored interval from the N. Smith Jr. #8 well is provided in Figures 9-17. The descriptions of the individual facies presented below are arranged from shallowest to deepest to facilitate easier reference to the core photographs. Facies A-F are labeled, and facies contacts are shown in the photographs to provide easy correlation to the sketched sequence for this core in Figure 8.

Facies F

Facies F (13,172-13,178 ft) is a lenticular-bedded and laminated mudstone (55% silt and 40% clay). A gradual decrease in the presence of glauconite from less than 3% at the base to less than 1% at the top of the unit is observed. Traces of pyrite (< 2%) are concentrated adjacent to and within silt-filled burrows. Siltstone laminae average 3-5 mm thick and contain both connected and disconnected lenses (Fig. 9). Burrow types within the interval are primarily horizontal forms, silt

Figure 9. Core photograph of 13,172–13,181 ft (4,015–4,018 m) from the N. Smith Jr. #8 well, showing the lowermost portions of facies F and uppermost portions of facies E. Lenticular bedding (a), glauconite pellets (b), and siderite concretions (c) are shown. Core is approximately 3 in. (7.5 cm) in diameter.

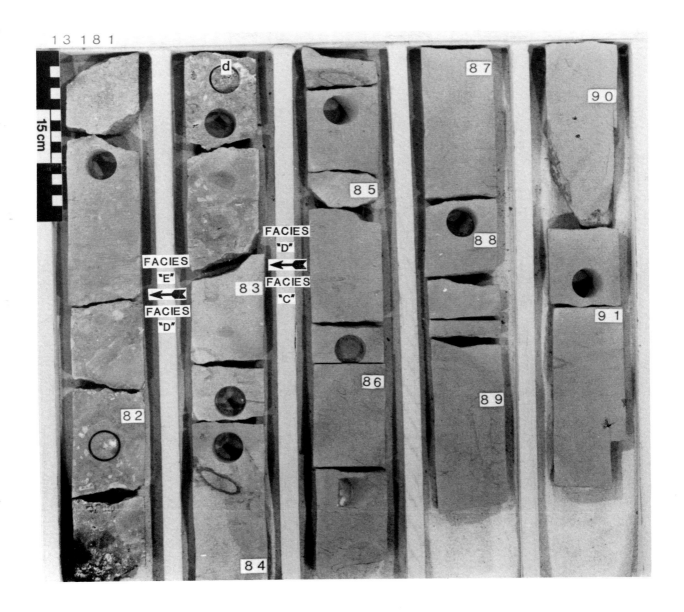

Figure 10. Core photograph of 13,181–13,192 ft (4,018–4,021 m) from the N. Smith Jr. #8 well. Note sharp contact (arrow) between facies C and facies D. Carbonate concretions (d) are common features in facies D.

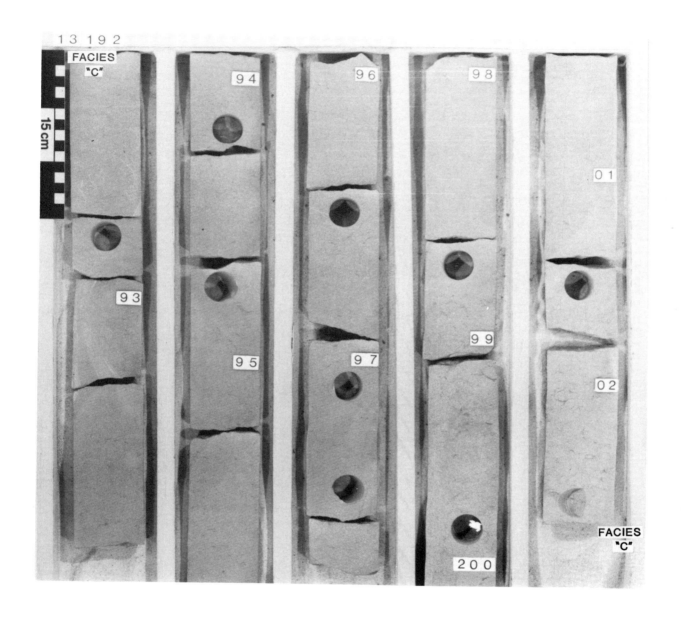

Figure 11. Core photograph of 13,192-13,203 ft (4,021-4,024 m) from the N. Smith Jr. #8 well showing the massive-appearing sandstones characteristic of facies C.

Figure 12. Core photograph of facies C from 13,203–13,213 ft (4,024–4,027 m) in the N. Smith Jr. #8 well. Note decrease in the amount of burrowing upward.

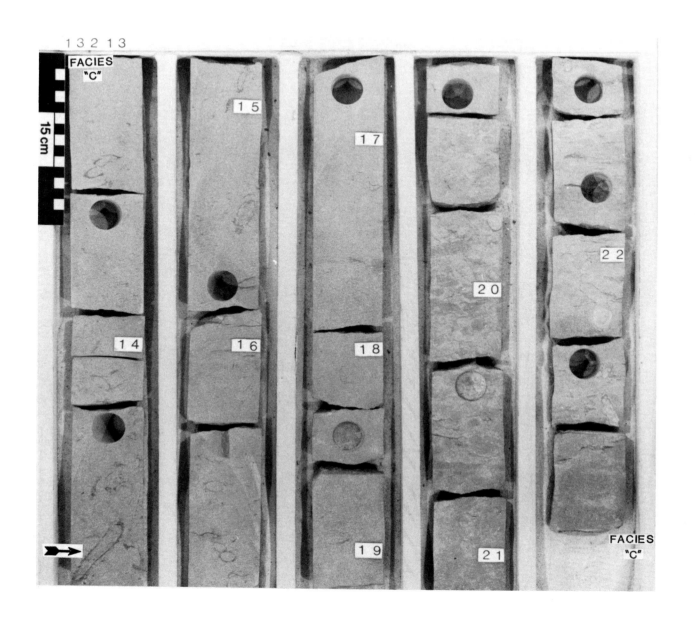

Figure 13. Core photograph of 13,213–13,223 ft (4,027–4,030 m) from the N. Smith Jr. #8 well. Note location of *Ophiomorpha* burrow at 13,214.5 ft (see Fig. 19A).

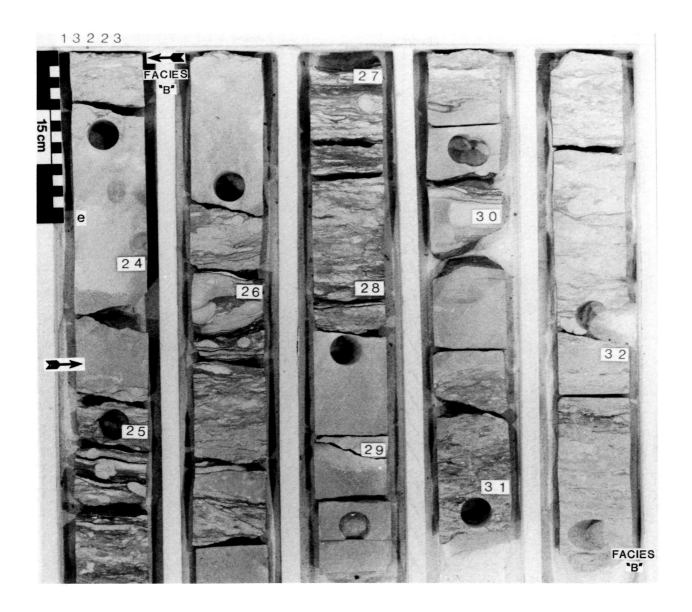

Figure 14. Core photograph of 13,223-13,233 ft (4,030-4,033 m) from the N. Smith Jr. #8 well showing characteristic features of facies B. Note amalgamation of sandstone beds (e) at 13,224 ft and discrete sandstone beds at 13,224.5 ft (see close-up photo in Fig. 20A).

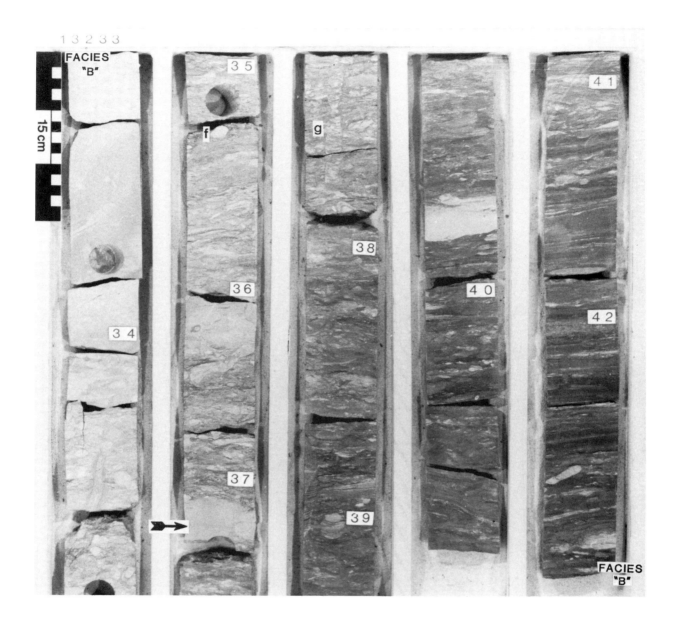

Figure 15. Core photograph of facies B from 13,233–13,243 ft (4,033–4,036 m) in the N. Smith Jr. #8 well. Note *Planolites* burrow (f), and *Teichichnus* burrow (g). Thin discrete sandstone bed at 13,237.2 ft is shown in close up photo in Figure 20C.

Figure 16. Core photograph of 13,243–13,250 ft (4,036–4,039 m) from the N. Smith Jr. #8 well. Note *Terebellina* burrow (h), and *Chondrites* burrow (i) in facies B. Load-casted ripple at 13,245.8 ft is shown in detail in Figure 21A.

Figure 17. Core photograph of facies A from 13,250–13,260 ft (4,039–4,042 m) in the N. Smith Jr. #8 well. Note contorted (j) and lenticular (k) bedding. Contorted bedding at 13,251.6 ft is shown in detail in Figure 21C.

Figure 18. Close-up photograph of calcareous sandstone from facies D (see Fig. 10 at 13,182 ft). Note abundant shell fragments.

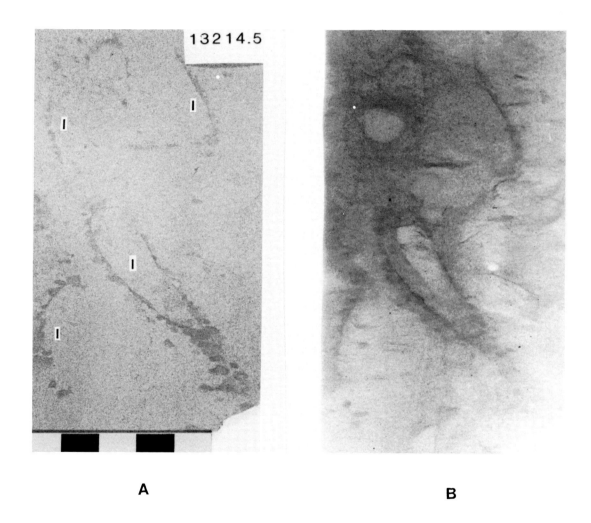

13214.5

A

B

Figure 19. [A] Close-up photograph of burrowed sandstone from facies C (see Fig. 13 at 13,214.5 ft). Note abundance of Ophiomorpha burrows (1). [B] X-ray radiograph of cored interval in Figure 19A. Note abundance of burrowing and lack of any preserved physical sedimentary structures.

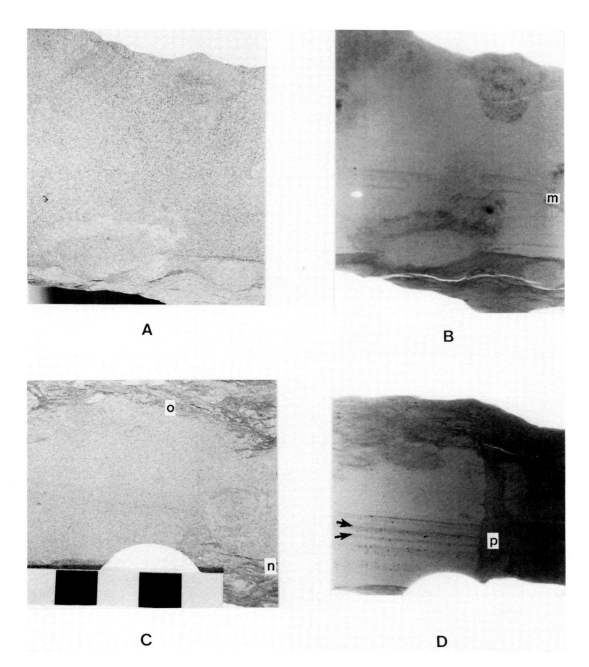

Figure 20. [A] Close-up photograph of thin (< 5 cm) very fine-grained sandstone bed in facies B (see Fig. 14 at 13,224.5 ft). [B] X-ray radiograph of cored interval in Figure 20A. Note faint horizontal to low-angle planar-tabular laminations at the base of the photo (m). [C] Close-up photograph of thin (< 5 cm) very fine-grained sandstone bed from facies B (see Fig. 15 at 13,237.2 ft). Note sharp lower contact (n) and burrowed upper contact (o). [D] X-ray radiograph of cored interval in Figure 20C. Note horizontal laminations and normal grading within laminations. Arrows represent coarser-grained (light) and finer-grained (dark) sediment. Also, note burrowing, which has subsequently destroyed laminations (p).

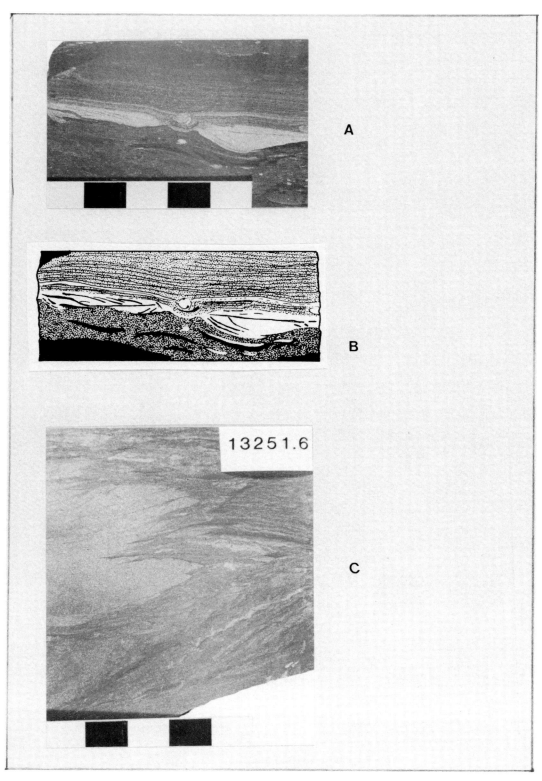

Figure 21. [A] Close-up photograph of a load-casted ripple from facies A (see Fig. 16 at 13,245.8 ft). [B] Sketch of cored interval in Figure 21A. Note truncation surfaces of individual laminations. [C] Close-up photograph of cored interval in facies A illustrating contorted bedding resulting from soft-sediment deformation (see Fig. 17 at 13,251.6 ft).

filled and ovate. Overall diversity and abundance of burrow types is low.

Facies E

Facies E (13,178–13,181.5 ft) is a glauconite-rich (10%–15%) bioturbated sandy mudstone. Individual glauconite pellets range in size from 3 to 6 mm in diameter. The sand fraction (mean grain size, 80–90 µ), constitutes 40% of the total interval and decreases in abundance upward to less than 10%. This unit is extensively burrowed to bioturbated (Figs. 9, 10). Rare horizontal laminations are the only physical sedimentary structures observed. Siderite concretions are also present in this facies. The uppermost contact is gradational over a 30–40 cm interval and corresponds to a gradual decrease in glauconite content. The lower contact is sharp.

Facies D

Facies D (13,281.5–13,283 ft), a calcareous, very fine-grained sandstone, is a massive-appearing, relatively thin unit (1.5 ft; 0.5 m) and exhibits grain-size characteristics similar to those of facies C (mean grain size, 100–125 µ; maximum, 150 µ); however, they differ mineralogically by the presence of abundant, diverse calcareous fragments and epigenetic carbonate concretions in facies D (Fig. 10). Individual foraminifers (*Miliolid*), and bivalve and gastropod fragments are present within this facies (Fig. 18). The localized abundance of carbonate material has yielded a tightly cemented interval overlying the main reservoir sandstone. No physical or biogenic sedimentary structures are observed in facies D.

Facies C

Facies C (13,183-13,223 ft), a burrowed and massive-appearing, very fine-grained sandstone, represents the major producing reservoir within the W8 Sandstone. Average thickness of facies C within the central part of the field is 40 ft (12.2 m). Relative proportions of sand-size to silt-size material range from 60% and 30%, respectively, near the base, to 80% and 15% throughout the remainder of the interval. Average grain size is 100-125 μ with a maximum of 150 μ. Traces of glauconite (< 2%) are also present. There is a general cleaning-upward trend within the facies as the mud content decreases upward.

The entire interval has been extensively burrowed, and therefore few physical sedimentary structures are preserved (Figs. 10, 11, 12, 13). Ophiomorpha burrows are the predominant burrow type and are most common in the lower half of the facies (Figs. 19A, B). In the upper half of the unit, fewer distinct burrows can be recognized, and the intensity of burrowing decreases.

Facies B

Facies B (13,223-13,245 ft), a burrowed muddy sandstone with laminated sandstone interbeds, is an interbedded unit that grades from a sandy mudstone at the base (20% sand) to a muddy, very fine-grained sandstone (70%-80% sand) at the upper contact. Well-sorted, very fine-grained sandstone (mean grain size, 90-100 μ) beds from 5 to 30 cm thick are interbedded with the burrowed muddy sandstone (Figs. 14, 15, 16). Glauconite (2%-3%) occurs throughout, with traces (< 1%) of pyrite occurring primarily in the basal portions of the unit.

The discrete sandstone beds within facies B exhibit sharp lower contacts (Figs. 14, 15, 20A, C). Individual bed thickness and frequency of occurrence of individual beds increases upward in the sequence (Fig. 14). The thickest of the sandstone beds (30 cm) occurs near the top of facies B and may represent amalgamation of smaller individual beds. Although many of the sandstone beds are massive in appearance (Fig. 20A, C), X-ray radiographs reveal horizontal to low-angle planar-tabular laminations near the base of each bed (Fig. 20B). Fine-scale, normal-graded bedding is also observed in the discrete sandstone beds of facies B (Fig. 20D).

Numerous burrows occur within facies B, including Teichichnus, *Planolites*, and *Terebellina* (Figs. 15, 16). The diversity of burrows is generally low to moderate, and the greatest diversity occurs within the muddy sandstone. It should also be noted that the discrete sandstone beds of facies B have the highest recorded permeability and porosity values (98 md and 25%, respectively), as determined from plug data (Table 1).

Facies A

Facies A (13,245-13,260 ft) is a mudstone (50% clay; 45% silt) characterized by the presence of contorted and lenticular beds (Figs. 16, 17). Lenticular siltstone beds are 1-10 mm thick, flat, and connected. Traces of glauconite (< 1%) occur adjacent to fragments of organic detritus, primarily in the lower portion of the interval. Contorted bedding is a product of soft-sediment deformation and exhibits components of thrusting and considerable disruption of the original bedding in facies A (Figs. 17, 21C). Approximately 40% of facies A exhibits this style of bedding. The most abundant physical sedimentary structures are thin (1-3 mm), streaky, and lenticular siltstone laminations. In a few places, these siltstone lenses may be recognized as load-casted ripples (Figs. 16, 21A, B). These sedimentary structures indicate a depositional environment where the hydraulic regime is constantly fluctuating between periods of increased and decreased fluid motion, and/or where sediment supply is episodic.

Biogenic activity, as recognized by the degree of burrowing in the facies, is relatively low (< 10% of the cored interval). Few distinct burrows can be recognized; however, those which do occur are primarily ovate, less than 1.5 cm in diameter, horizontal, and siltstone filled.

Distinction of the upper contact between facies A and B (Fig. 16) is gradational and based on the decrease in the amount of soft-sediment deformation observed at the top of facies A and the increase in the degree of burrowed siltstone laminations at the base of facies B.

Depositional Processes and Events

Evidence from physical and biogenic sedimentary structures observed in the cored interval suggest three major depositional events: (1) initial deposition during progradation of inner-shelf and shoreface sediments over outer-shelf/upper-slope muds (facies A-C); (2) erosion of upper-shoreface and foreshore sediments during the subsequent transgression and deposition of a transgressive lag as a ravinement surface (facies D); and (3) continued sea level rise and deposition of shelf muds below storm-weather wave base (facies E and F).

Data suggest that facies A was deposited in an outer-shelf to upper-slope setting. While dominated by deposition from suspension of muds, this environment was punctuated with periods of increased energy and sediment (i.e., silt and sand) availability. The numerous siltstone laminae in facies A are believed to represent the distal portions of the discrete sandstone beds common in facies B. The occurrence of discontinuous siltstone lenses reflects the development of small-scale ripple bedding in an otherwise sediment-deficient environment. The frequent occurrence of contorted bedding in facies A appears to be a product of deposition on an unstable substratum, subject to gravity-flow processes. Another important aspect of this facies is the lack of biogenic sedimentary structures. This suggests either that sedimentation rates were extremely high or that the environment was hostile to benthic organisms. If the sedimentation rate was so rapid as to be environmentally stressful for burrowing organisms, then it would appear unlikely that such a diverse and abundant ichnofauna would occur in the overlying facies. Since the mudstones in the overlying facies (B) are extensively burrowed, we conclude that facies A was deposited under oxygen-deficient conditions (see Anisgard, 1970). The common occurrence of pyrite nodules in facies A also suggests deposition under reducing conditions. Increasing biogenic activity in the overlying facies is in response to aggradation of the seafloor and the development of more oxygenated marine conditions.

In addition to the increase in abundance and diversity of burrow types in facies B, there is a concomitant increase in the frequency and intensity of episodes of coarser-grained (fine-sand-size) sedimentation. Sedimentary structures and sequences within the discrete sandstone beds are similar to those observed by Brenchley (1985). The sharp basal contact in each bed is believed to form in response to initial erosion by basinward-directed geostrophic flows (Swift et al., 1985), and the burrowed upper contact during postdepositional fair-weather periods. With the exception of horizontal parallel laminations and minor normal-graded bedding, none of the beds examined contained the idealized tempestite sequences described by Bourgeois (1980), Aigner (1985), or Walker (1984). It is possible that extensive burrowing at the upper contact destroyed any manifestation of hummocky or wave-ripple cross-stratification. However, it is worth noting that the dimensions of hummocks (3-10 ft; 1-3 m) are such that direct observation of this stratification type in 3-in. (7.5-cm) cores is extremely difficult. Perhaps only indirect evidence, such as low-angle truncation surfaces, multiple directions of dip of laminasets, draping of laminae, and erosional lower contacts may be used to infer the presence of this feature in conventional cores. On the basis of trace fossil assemblage, high abundance and diversity of burrow types, and the increasing frequency and upward thickening of sandstone beds, this facies is interpreted to have formed in a zone seaward of fair-weather wave base, but within storm wave base.

Facies C reflects deposition in an environment dominated by biogenic activity. There is, however, a subtle yet significant change in the relative rate of physical versus biogenic processes within the facies. In the lowermost half of the interval, well-developed Ophiomorpha burrows are abundant, and mud is dispersed throughout the sandstone matrix. Although grain size remains essentially unchanged, the degree of burrowing and mud content decreases, reflecting an increasing influence of physical processes (waves). Within the upper part of facies C, localized horizontal planar-tabular laminations are found. These parallel laminations probably formed by deposition from suspension in response to wave action, suggesting an aggradation of the facies into a more energetic zone. The degree of wave action, however, was evidently insufficient to dominate over biogenic processes.

Facies C was deposited in the transition zone and lower-shoreface environments as defined by Reineck and Singh (1980) and Howard and Reineck (1981). Problematic in this interpretation is the fact that facies C is far thicker than the entire beach-to-shelf sequence observed in modern low-wave-energy environments. However, for higher-wave-energy environments, Howard and Reineck (1981) have shown that, although sedimentary sequences and characteristics remain the same, the entire shoreface and transition-zone package is substantially thicker. This unusual thickness (> 12 m) is evident in facies C and reflects the higher wave energies associated with deposition at the shelf margin.

The contact between facies C and D is a disconformity in the cored sequence and marks the initiation of a transgressive phase of deposition. The hiatus is reflected in the lack of upper-shoreface and foreshore stratification in the cored sequence. The abundance of carbonate shell fragments and sand-size grains in facies D is associated with decreased sediment supply and a winnowing of fine-grained material.

Facies E is glauconite-rich and bioturbated, reflecting the dominance of biogenic processes. This facies represents deposition on a sediment-deficient shelf, probably below storm wave base, during a transgression.

With the continued rise of relative sea level, facies F was deposited in deeper water on the shelf under conditions hostile to bottom-dwelling fauna. Although similar to facies A, this facies lacks the contorted bedding prevalent in facies A. Laminated siltstone lenses suggest episodic coarser-grained sedimentation. Facies F represents the culmination of the depositional event responsible for the formation of the W8 Sandstone.

Depositional Model

The geometry, lateral continuity, thickness of the sequence, and proximity to the shelf margin suggest that the W8 Sandstone represents a progradational shoreface sequence that formed at or near the shelf edge (Fig. 22). While the role of subsidence due to growth faulting or sediment compaction cannot be dismissed as a possible mechanism creating such a thick lower-shoreface sequence, it is equally likely that the thickness of the sequence reflects primary depositional control.

Sand bodies forming at a shelf margin are subject to much higher wave energy than those forming on a broad shallow shelf, where wave refraction and energy dissipation are greater. The two sandstone intervals underlying the W8 Sandstone (Fig. 4) probably represent similar depositional events, so that the depositional episode is made up of three stacked prograding shoreface sequences. Since there is no direct evidence of a fluvial system associated with the sand body, the applicability of a shelf-edge delta model is debatable. Although it is implicit in the development of a progradational sequence at the shelf edge that there must also be an associated fluvial system (Suter and Berryhill, 1985), the recognition of these features may be beyond the resolution afforded by the available subsurface data.

Reservoir Quality

Of the six sedimentary facies recognized in the Fordoche Field study area, only two (B and C) are associated with any significant hydrocarbon production. The uppermost half of the burrowed and

WILCOX SHELF-EDGE DELTA MODEL

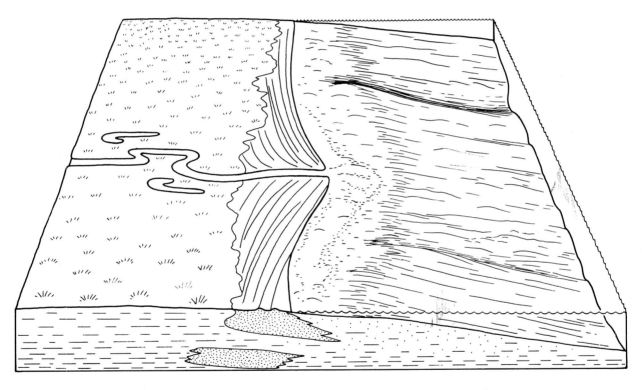

Figure 22. Block diagram depicting a paleogeographic reconstruction of the depositional setting for the W8 Sandstone, specifically a prograding shoreface system at the shelf edge.

massive-appearing sandstone of facies C and the laminated sandstone interbeds of facies B display the highest permeability and porosity measurements as determined from core-plug data (Table 1) and hence possess the best "reservoir quality."

The laminated sandstone interbeds of facies B have the highest average permeability (10-98 md) and porosity (10%-23%) values in the W8 Sandstone (Fig. 23). Average oil saturation values for these beds is 15%-23%. However, although these values are strongly suggestive of high reservoir quality, the sandstone beds of facies B are relatively thin, laterally discontinous, and interbedded with nonpermeable mudstones (Figs. 8, 23). Therefore, despite the intermittent high porosity and permeability values, facies B lacks reservoir homogeneity and continuity and is classified as relatively poor quality.

Facies C, burrowed and massive-appearing, very fine-grained sandstone, has the highest overall reservoir quality and is the main producing interval within the W8 Sandstone. Measured porosity values in facies C show minimal variation in the cored interval of the N. Smith Jr. #8 and range from 15% at the base of the unit to a maximum of 22% at the top (Fig. 23). Permeability, and overall reservoir quality for facies C, seems strongly controlled by the original depositional fabric of the sandstone, as there is a strong correlation between permeability trends and the degree of biogenic versus physical sedimentary structures. The lower two-thirds of facies C (13,195-13,223 ft) is a highly burrowed to bioturbated interval with average permeability values of 0.1-1.0 md (Figs. 8, 23). It seems likely that the high degree of biogenic reworking has altered the original depositional fabric of the sandstone. The silt and clay linings of the burrowed traces in this interval

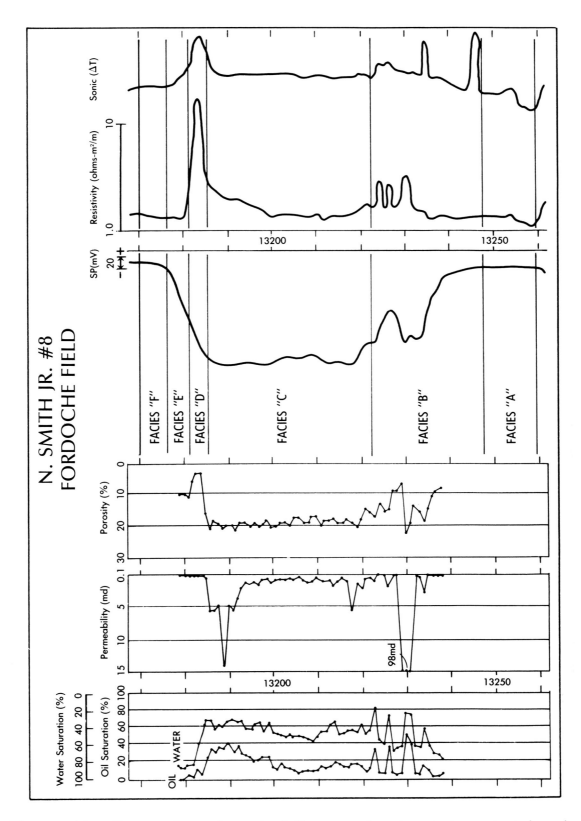

Figure 23. Reservoir characteristics and downhole electric log signatures for cored sedimentary facies from the W8 Sandstone interval in the N. Smith Jr. #8 well. Petrophysical data is from core plugs. Note extremely high permeability at the top of facies C and in the thin beds at 13,230 ft in facies B.

probably yield numerous small-scale permeability barriers. The upper one-third of facies C (13,183-13,195 ft) is a massive-appearing sandstone having low-angle planar laminations and relatively few burrow traces. Average permeability values in this interval are 5-15 md (Fig. 23). This high degree of reservoir quality is apparently due to the lack of biogenic reworking. The reservoir quality of facies C is greatly enhanced by its consistency in thickness and lateral continuity within Fordoche Field (Fig. 24), which yields a homogenous and continuous (along-strike) reservoir.

Another important facies within the W8 Sandstone interval is facies D, a well-cemented, calcareous, very fine-grained sandstone. Average measured porosity values of 4% observed in facies D are probably a product of the dissolution of shell clasts in the sandstone matrix. Calcareous cementation is responsible for the very low average permeability values (< 0.1 md) (Table 1; Fig. 23). There is a high probability that facies D, a very poor-quality reservoir unit, is instead a very important diagenetic seal for hydrocarbons. Facies D immediately overlies the main reservoir unit (facies C) and is laterally continuous throughout most of Fordoche Field (Fig. 24), which adds to its potential value as a stratigraphic seal.

Conclusions

1.   The downdip uppermost Wilcox trend in south-central Louisiana including Fordoche Field is coincident with the location of the lower Cretaceous carbonate reef trend and upper Cretaceous clastic shelf margins. The position of the shelf edge was maintained at least through deposition of the uppermost Wilcox.

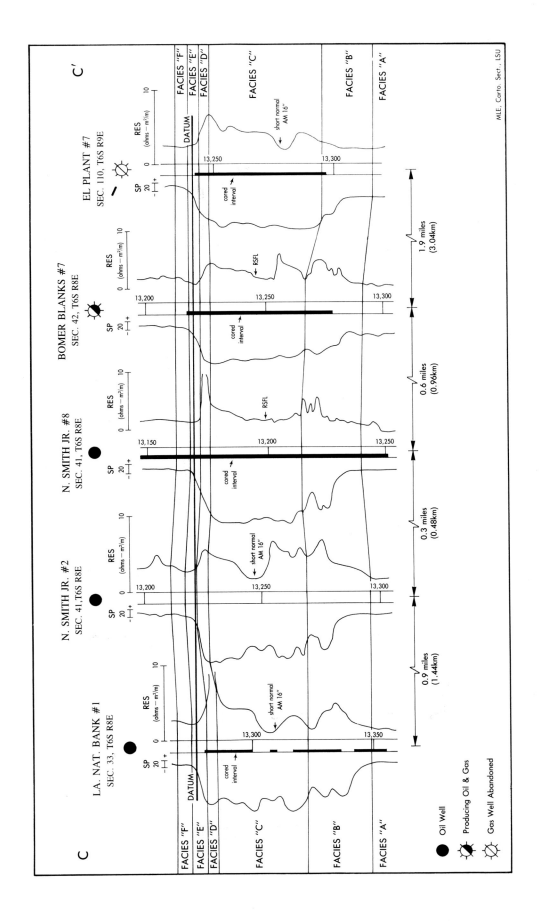

Figure 24. Strike-oriented cross section (C–C') within Fordoche Field (see Fig. 7 for location of the section). Cored intervals in each well are shown by solid black bars.

405

2. Fordoche Field is an important Wilcox oil and gas field that produces within a strike-oriented exploration trend that extends east to west in south-central Louisiana and is interpreted as a paleo shelf margin. Additional production should be found along this downdip trend.

3. On the basis of electric-log correlations and sedimentologic analysis of conventional cores, we propose that deposition of the W8 Sandstone occurred as a prograding shoreface at or near the shelf edge.

4. Shoreface deposits constitute the main reservoir facies, although production from tempestites underlying the main sandstone may also be important. Many of these storm beds have the largest permeability and porosity values of the entire sequence; however, their lack of lateral continuity and high degree of burrowing, which produces numerous permeability barriers, make these beds poor reservoirs, requiring the use of enhanced-recovery methods.

Acknowledgments

The authors are deeply indebted to the management and personnel of the Sun Oil Company, especially Ms. Judy Melvin, Mr. Mark Gaby, and Mr. Charles Mayne who provided petrophysical data, well logs and cores from Fordoche Field. Without their kind assistance this study could not have been conducted. The Sun Oil Company is also thanked for providing its permission to utilize the N. Smith Jr. #8 core for this workshop.

The Atlantic Ritchfield Company is gratefully acknowledged for financial support to the senior author in the form of a fellowship.

This manuscript is part of an in-depth regional Wilcox study being jointly conducted by the Basin Research Institute and Louisiana Geological Survey. Tory Eddins provided detailed information on the production history and reservoir engineering parameters of the W8 Sandstone. Lori Nunn analyzed foraminiferal assemblages from mudstones overlying the W8 Sandstone cored interval. The directors of the Basin Research Institute (C. H. Moore) and the Louisiana Geological Survey (C. G. Groat) are thanked for their continuous support and encouragement in this and other subsurface studies of the northermn Gulf of Mexico.

References

Aigner, T., 1985, Storm depositional systems, in Freidman, G. M., Neugebauer, H. J., and Seilacher, A., eds., Lecture notes in earth sciences: New York, Springer-Verlag, v. 3.

Albach, D. C., 1979, The depositional history of the uppermost Wilcox (lower Eocene) of west-central Beauregard Parish, Louisiana: Baton Rouge, Louisiana State University, Master's thesis.

Anisgard, H. W., 1970, Causes of dominantly arenaceous foraminiferal assemblages in downdip Wilcox of Louisiana: Gulf Coast Assoc. Geol. Socs. Trans., v. 20, p. 210-17.

Berg, R. R., and Tedford, F. J., 1977, Characteristics of Wilcox gas reservoirs, northeast Thompsonville Field, Jim Hogg and Webb Counties, Texas: Gulf Coast Assoc. Geol. Socs. Trans., v. 27, p. 6-19.

Bourgeois, J., 1980, A transgressive shelf sequence exhibiting hummocky stratification--the Cape Sebastian Sandstone (upper Cretaceous), southwestern Oregon: Jour. Sed. Petrol., v. 50, p. 681-702.

Brenchley, P. J., 1985, Storm influenced sandstone beds: Modern Geol., v. 9, p. 369-96.

Busch, D. A., 1975, Influence of growth faulting on sedimentation and prospect evaluation: Amer. Assoc. Petrol. Geol. Bull., v. 59, p. 217-30.

Christina, C. C., and Martin, K. G., 1979, The lower Tuscaloosa trend of south-central Louisiana, "You ain't seen nothing till you've seen the Tuscaloosa": Gulf Coast Assoc. Geol. Socs. Trans., v. 29, p. 37-41.

Chuber, S., and Begeman, R. L., 1982, Productive lower Wilcox stratigraphic traps from an entrenched valley in Kinkler Field, Lavaca County, Texas: Gulf Coast Assoc. Geol. Socs. Trans., v. 32, p. 255-62.

Coates, E. J., 1979, The occurrence of Wilcox lignite in west-central Louisiana: Baton Rouge, Louisiana State University, Master's thesis.

Craft, W. E., 1966, Channel sands are the key to Wilcox oil: Oil and Gas Jour., v. 64, no. 15, p. 124-30.

Durham, C. O., and Peoples, E. M., 1956, Pleistocene fault zone in southeastern Louisiana: Gulf Coast Assoc. Geol. Socs. Trans., v. 6, p. 65-66.

Eckles, W. W., Jr., Prihoda, C., and Holden, W. W., 1981, Unique enhanced oil and gas recovery for very high-pressure Wilcox sands uses cryogenic nitrogen and methane mixture: Jour. Petrol. Tech., June, p. 971-83.

Edwards, M. B., 1980, The Live Oak delta complex--an unstable shelf-edge delta in the deep Wilcox trend of south Texas: Gulf Coast Assoc. Geol. Socs. Trans., v. 30, p. 71-79.

Edwards, M. B., 1981, Upper Wilcox Rosita delta system of south Texas: growth-faulted shelf-edge deltas: Amer. Assoc. Petrol. Geol. Bull., v. 65, p. 54-73.

Fisher, W. L., and McGowen, J. H., 1967, Depositional systems in the Wilcox Group of Texas and their relationship to occurrence of oil and gas: Gulf Coast Assoc. Geol. Socs. Trans., v. 17, p. 105-25.

Galloway, W. E., 1968, Depositional systems of the lower Wilcox Group, north-central Gulf Coast Basin: Gulf Coast Assoc. Geol. Socs. Trans., v. 18, p. 275-89.

Hendricks, L., and Wilson, W. F., 1967, Introduction, in Hendricks, L., ed., Commanchean (lower Cretaceous) stratigraphy and paleontology of Texas: Permian Basin section, Soc. Econ. Pal. Min. pub. 67-8, p. 1-6.

Howard, J. D., and Reineck, H. E., 1981, Depositional facies of high-energy beach-to-offshore sequence--comparison with low-energy sequence: Amer. Assoc. Petrol. Geol. Bull., v. 65, p. 807-30.

Hoyt, W. V., 1959, Erosional channel in the middle Wilcox near Yoakum, Lavaca County, Texas: Gulf Coast Assoc. Geol. Socs. Trans., v. 9, p. 41-50.

Jackson, M. P. A., and Galloway, W. E., 1984, Structural and depositional styles of Gulf Coast Tertiary continental margins, application to hydrocarbon exploration: Amer. Assoc. Petrol. Geol. Note Series no. 25, p. 1-226.

Lehner, P., 1969, Salt tectonics and Pleistocene stratigraphy on continental slope of northern Gulf of Mexico: Amer. Assoc. Petrol. Geol. Bull., v. 53, p. 2431-79.

Louisiana Geological Survey, 1981, Oil and gas maps of Louisiana: scale 1:380,160, 2 sheets.

Lowry, P., Lemoine, R. C., and Moslow, T. F., in press, Shelf-margin sedimentation in the Wilcox Group, south-central Louisiana: Gulf Coast Assoc. Geol. Socs. Trans.

Martin, R. G., 1977, Northern and Eastern Gulf of Mexico continental margin: stratigraphic and structural framework, in Bouma, A. H., Moore, G. T., and Coleman, J. M., eds., Framework, facies, and oil-trapping characteristics of the upper continental margin: Amer. Assoc. Petrol. Geol. Studies in Geology, no. 7, p. 21-42.

McCulloh, R. P., and Eversull, L. G., in press, Shale-filled channel system in the Wilcox, north-central south Louisiana: Gulf Coast Assoc. Geol. Socs. Trans.

Mulcahy, S. A., 1981, Near-surface lignite occurrence, Sabine Uplift, northwest Louisiana: Baton Rouge, Louisiana State University, Master's thesis.

Pierson, J. R., 1970, Fordoche Field, Pointe Coupee Parish, Louisiana, in Typical oil and gas fields of southeast Louisiana, v. 2, p. 9-9g.

Reineck, H. E., and Singh, I. B., 1980, Depositional sedimentary environments (2d ed.): New York, Springer-Verlag, 549 p.

Rogers, J. D., 1983, The occurrence of deep basin lignite in the Wilcox Group of northeast Louisiana: Baton Rouge, Louisiana State University, Master's thesis.

Self, G. A., Breard, S. Q., Rael, H. P., Stein, J. A., Traugott, M. O., and Eason, W. D., 1985, Lockhart Crossing Field, New Wilcox trend in southeastern Louisiana: Amer. Assoc. Petrol. Geol. Bull., v. 69, no. 5, p. 306.

Stehli, F. G., Creath, W. B., Upshaw, C. F., and Forgotson, J. J., Jr., 1972, Depositional history of Gulfian Cretaceous of East Texas Embayment: Amer. Assoc. Petrol. Geol. Bull., v. 69, p. 77-91.

Suter, J. R., and Berryhill, H. L., Jr., 1985, Late Quaternary shelf-margin deltas, northwest Gulf of Mexico: Amer. Assoc. Petrol. Geol. Bull., v. 69, p. 77-91.

Swift, D. J. P., Niederoda, A. W., Vincent, C. E., and Hopkins, T. S., 1985, Barrier island evolution, middle Atlantic shelf, U.S.A. Part I--shoreface dynamics: Marine Geol., v. 63, p. 331-61.

Thorsen, C. E., 1963, Age of growth faulting in southwest Louisiana: Gulf Coast Assoc. Geol. Socs. Trans., v. 13, p. 103-10.

Walker, R. G., 1984, Shelf and shallow marine sands, in Walker, R. G., ed., facies models: Geoscience Canada, reprint series 1, p. 141-70.

Winker, C. D., 1982, Cenozoic shelf margins, northwestern Gulf of Mexican Basin: Gulf Coast Assoc. Geol. Socs. Trans., v. 32, p. 427-48.

Winker, C. D., and Edwards, M. B., 1983, Unstable progradational clastic shelf margins, in Stanley, D. J., and Moore, G. T., eds., The shelfbreak--critical interface on continental margins: Soc. Econ. Pal. Min. Spec. Publ. No. 33, p. 139-57.

DEPOSITIONAL FACIES OF THE

MIDDLE TRIASSIC HALFWAY FORMATION,

WESTERN CANADA BASIN

Clarence V. Campbell and John C. Horne

RPI Canada Ltd., 1616, 540 – 5th Avenue S.W., Calgary, Alberta, Canada T2P 0M2

RPI International, 2845 Wilderness Place, Boulder, Colorado 80301

Abstract

The Middle Triassic Halfway Formation is a significant producer of oil and gas in the eastern British Columbia and western Alberta portions of the Western Canada Basin. This formation produces hydrocarbons primarily from stratigraphic traps.

The Halfway Formation was deposited on the northeastern shoreline of an epicontinental seaway that occupied much of western Canada during Middle Triassic time. The primary locus of sediment influx to the basin was located in northeastern British Columbia. From that area, sediment was transported eastward and southeastward under the influence of longshore currents set up by waves approaching from the west. During periods of shoreline progradation, wave-dominated deltas formed in eastern British Columbia, and barrier/strand-plain deposits accumulated downdrift in the western portion of Alberta. These barrier/strand-plain deposits were backed by arid, hypersaline lagoons, tidal flats, and sabkhas.

Two general stratigraphic sequences are recognized within the Halfway shoreline sandstones: (1) an upward-coarsening sequence consisting of offshore marine shales and siltstones, overlain by cross-bedded shoreface sandstones and capped by foreshore flat-bedded sandstones; and (2) a coarse-grained, tidal inlet-fill sequence consisting of flat-bedded to cross-bedded sandstone and shell-hash conglomerates.

In barrier island areas, the upward-coarsening sequence predominates. Because waves reworked the sediments deposited above the fair-weather wave base, the upper shoreface/foreshore sandstones originally had good primary intergranular porosity and permeability. However, these upper shoreface/foreshore deposits are generally thin (less than 5 m), and later diagenetic events have greatly reduced the primary intergranular porosity. These sandstone bodies are most continuous along depositional strike (northwest-southeast).

In some areas, inlets have scoured through the shoreface deposits, and the inlet-fill sequences, which are up to 20 m thick, commonly rest upon offshore marine siltstones and shales. Tidal currents moving through the inlets transported and deposited sand-size sediment as well as large amounts of shell material. As a result, these inlet-fill deposits had excellent primary intragranular porosity and permeability that was subsequently enhanced by the leaching of the calcareous shell material. Leaching resulted in the development of additional secondary moldic porosity and permeability. Among the Halfway deposits, the inlet-fill sequences have the greatest reservoir potential. They are laterally discontinuous along depositional strike (northwest-southeast) and more continuous along depositional dip (northeast-southwest).

Introduction

Triassic strata exposed in the foothills of eastern British Columbia and western Alberta, as well as in the subsurface of the Western Canada Basin, contain a perplexing variety of rock types and depositional settings ranging from offshore marine shales to shoreline sandstones and hypersaline coastal sabkha evaporites. Because of their heterogeneity, early attempts to correlate these strata using principles of biostratigraphy as well as lithostratigraphy met with little success.

Early drilling in the Western Canada Basin revealed hydrocarbons in anticlinal traps of Triassic strata. By the 1960s, most of the obvious structural traps had been tested. The discovery and subsequent development of prolific stratigraphic traps in the Milligan, PeeJay, and Wembley fields during the 1980s sparked new interest in these Triassic deposits. There was no general consensus among geologists on the underlying geological causes for the Triassic productive trends observed in the Western Canada Basin, and consequently, sound and widely accepted depositional models were lacking. Furthermore, diagenetic reduction of porosity of these strata had rendered them too tight to be of reservoir quality in many areas, and many Triassic production trends have been relegated to a role as secondary objectives.

The study area includes that part of the Western Canada Basin lying between Townships 61 and 87 and between Ranges 1 and 13 west of the sixth meridian (Fig. 1). The area includes a large portion of

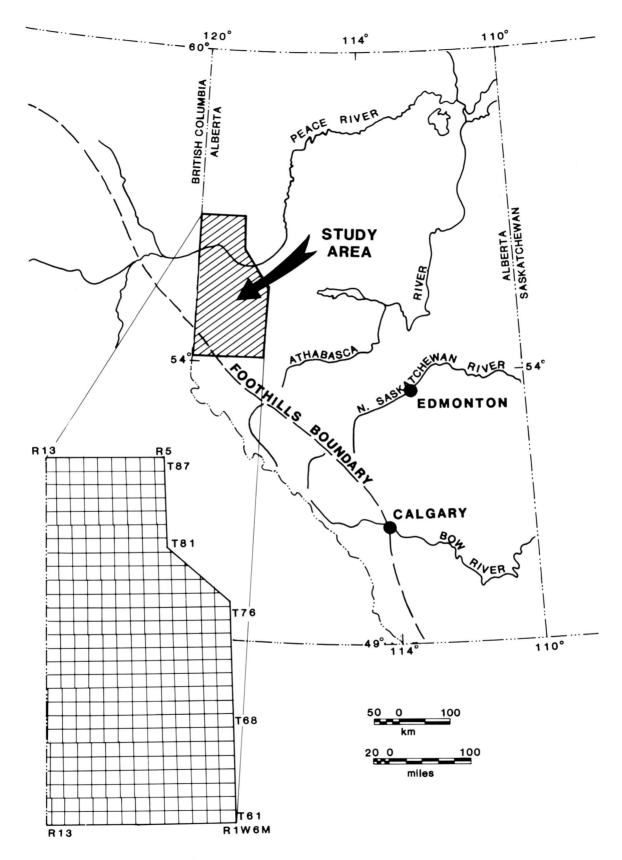

Figure 1. Location map of the study area.

the present Halfway hydrocarbon production within the Western Canada Basin. It encompasses a total area of approximately 25,000 km² from which a total of 230 cores and logs from 936 wells were examined to assess the depositional patterns within the Halfway.

Stratigraphy

General

Previous studies of the Middle Triassic strata (Fig. 2) in the Western Canada Basin in Alberta have emphasized the differences between facies and depositional patterns of the Doig and Halfway Formations. An unconformity representing a substantial period of nondeposition or erosion is commonly inferred to exist between them in the Western Canada Basin (Armitage, 1962). Biostratigraphic evidence confirms the existence of such an unconformity (Chunta, 1969). Stratigraphic evidence presented in this study supports this conclusion but indicates that the unconformity is not at the stratigraphic position identified by previous studies.

Doig Formation

The Doig Formation, named by Armitage (1962), constitutes the upper part of the Daiber Group. It was described from a stratigraphic section in the subsurface (Texaco N. F.A. Buick Creek No. 7 location-- Lsd. 6, Sec. 26, Twp. 87, Range 21, W6 Meridian). The lower contact of the formation is placed at the base of the phosphate-rich beds that overlie the transgressive disconformity at the top of the Montney Formation. Doig strata above these prominent phosphate-rich beds consist mostly of dark grey to black siltstones and shales. The upper contact of the Doig is at the top of a thin sandstone associated with another

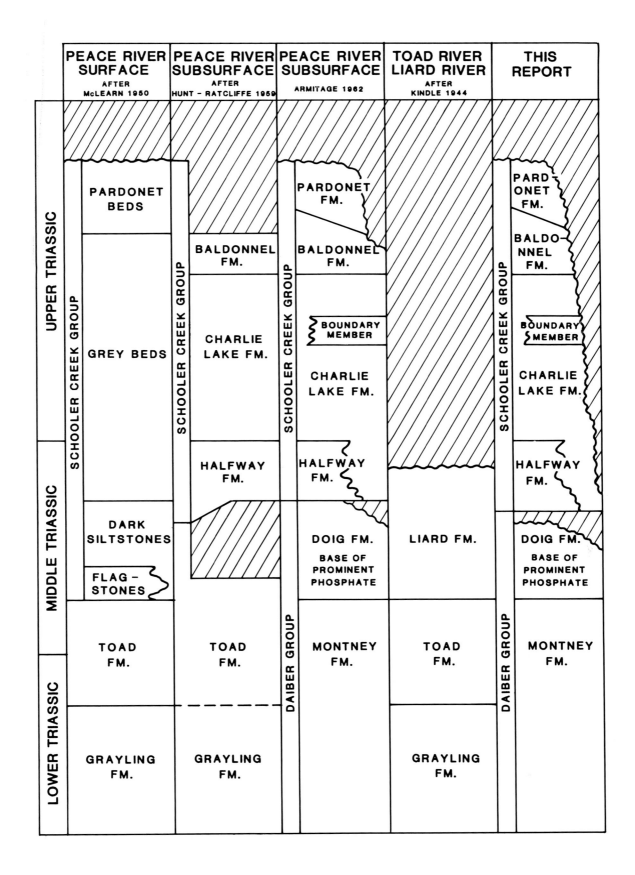

Figure 2. Stratigraphic nomenclature table.

transgressive disconformity. The Doig Formation is more than 120 m thick in the rapidly subsiding western and southwestern portions of the study area and pinches out along the northeastern margin. Some of the thinning is depositional and related to the slower rate of subsistence along the northeastern margin of the Western Canada Basin. However, the majority of the thinning is the result of subaerial erosion during a sea level low stand prior to a subsequent sea level rise and transgression at the end of the Doig deposition.

Halfway Formation

The Halfway Formation constitutes the lower part of the Schooler Creek Group in the Western Canada Basin (Hunt and Ratcliffe, 1959). Unlike previous studies (Armitage, 1962; Gibson, 1972), which placed the lower contact of the Halfway Formation at the base of the Halfway sandstone, this paper places the boundary on the transgressive disconformity at the top of the Doig Formation on the basis of core analysis. Cores penetrating through the transgressive lag generally consist of shell hash, pebble conglomerates, and massive to burrowed sandstones. In areas of the underlying Doig channels (Wembley and Sinclair fields), the boundary is comparatively easy to pick. A marked density increase on petrophysical logs is commonly associated with the transgressive disconformity at the Doig Formation. In areas where the lag is either nonexistent or very thin, the contact becomes more difficult to pick but may be inferred from surrounding wells with well-developed lag deposits.

Placing the boundary at the transgressive lag rather than at the base of the Halfway sandstone is important because of the genetic relationships of the units. The shales overlying the lag are associated with

418

the Halfway Formation and are interpreted to be offshore components of the Halfway shoreline sands. The sandstone units below the transgressive lag were deposited prior to those of the Halfway and have different depositional and diagenetic characteristics.

The upper contact of the Halfway Formation is a facies boundary between the sandstones of the Halfway and the red-bed clastics and evaporites of the Charlie Lake Formation. In most places, this boundary is abrupt, but the contact rises stratigraphically to the west. The Halfway Formation is more than 85 m thick in the southwestern portion of the Western Canada Basin and pinches out at the eastern edge of the basin where it is truncated by the pre-Nordegg unconformity of Jurassic age.

As recognized by Armitage (1962), the upper portion of the Halfway Formation interfingers laterally and is time equivalent with the lower portion of the Charlie Lake Formation. This interfingering of the lower Charlie Lake strata to the west with the Halfway sandstones occurred in response to a rise in sea level in the Triassic seaway.

The initial Halfway deposits overlying the transgressive disconformity at the top of the Doig Formation comprise marine shales and siltstones. These shales and siltstones are thickest in the western portion of the study area and thin toward the northeastern edge of the basin. In some places these marine siltstones and shales coarsen upward into the overlying Halfway sandstone, while in other areas, an abrupt scoured lower contact exists between the Halfway sandstone and the underlying marine siltstones and shales. In general, the Halfway sandstones become thinner and finer grained to the east and southeast. However, there are two trends to the thickness patterns of the Halfway sandstone. The coarsening-upward sandstone bodies are elongate parallel to the trend of the basin margin (northwest-southeast). By contrast, the sandstone units that have scoured bases and contain abundant amounts of transported shell material are isolated bodies that display thickness trends oriented in a northeast to southwest direction.

Charlie Lake Formation

The Charlie Lake Formation constitutes the main portion of the Schooler Creek Group in the Western Canada Basin (Hunt and Ratcliffe, 1959; Armitage, 1962). This study examines only the portion of the Charlie Lake Formation that laterally interfingers with and directly overlies the Halfway Formation. The Charlie Lake Formation consists primarily of red-bed clastic siltstones, sandstones, and shales, as well as evaporites, primarily anhydrites. Some of these evaporite beds are laterally continuous and provide excellent correlation markers within the Charlie Lake Formation. These evaporites are most continuous laterally parallel to the basin margin (northwest-southeast).

Depositional Setting

Regional Deposition

The distribution of Halfway sediments is best understood in the context of the regional depositional patterns that existed in the Middle Triassic seaway of western Canada. While rifting and sea-floor spreading were separating Europe and Africa from North America and numerous tensional block-faulted basins were developing along the continental margins, much of British Columbia and western Alberta was covered by a shallow epicontinental sea (Fig. 3), which spread over a tectonically stable platform. Sediments accumulating in this area onlapped Paleozoic

MIDDLE TRIASSIC PALEOGEOGRAPHIC RECONSTRUCTION

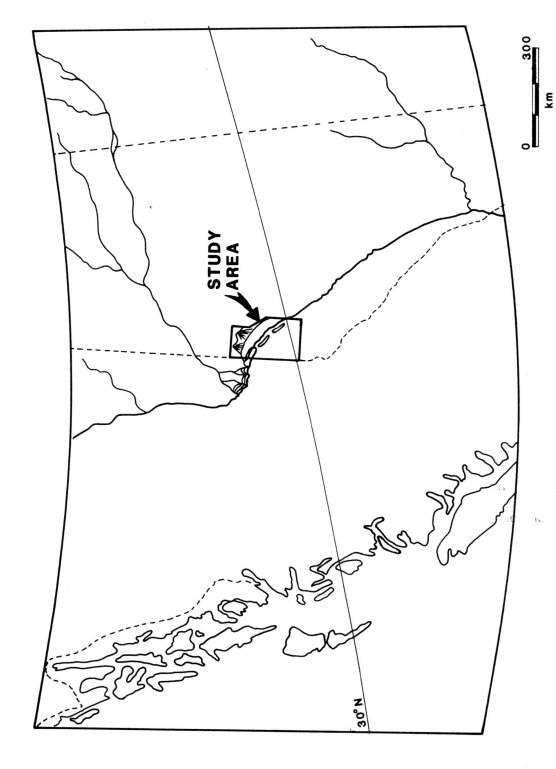

STUDY AREA

30° N

0   300
km

Figure 3.   Paleogeographic reconstruction of the Middle Triassic.

rocks that covered the Canadian Shield. Low-relief source areas to the north and east, including the Canadian Shield, supplied sediment to the northeastern shoreline of this seaway (Barss et al., 1964). These source areas provided sediment to the depositional basin from the erosion of Paleozoic terrigenous clastic and carbonate rocks. Because these sediments are second-cycle sedimentary deposits, the sandstones of the Middle Triassic are mineralogically and texturally mature. The shallow epicontinental sea advanced and retreated from the Pacific area a number of times during the Triassic (Peterson et al., 1973), creating a complex depositional and erosional history along the northeastern margin of the Western Canada Basin.

Basin Configuration

The Triassic Period is represented by fewer marine deposits in North America than any other geologic period (Peterson et al., 1973). In North America, rock types of the Triassic are typically continental and consist of red and green sandstones, shales, interbedded sandstones and shales, evaporites, and volcanics. Early Triassic deposition occurred over a wide belt in the western United States, but in western Canada was limited to the Western Canada Basin. Middle and Late Triassic deposition was not widely distributed in the western United States; however, in western Canada, such deposition continued in the eugeosynclinal area, as well as in the Western Canada Basin.

The Western Canada Basin is a foreland basin that developed during Triassic time. Within the study area, the structure is a fairly simple monocline dipping to the southwest into the more rapidly subsiding portion of the foreland basin. Structural dips vary from about 0.40° (7 m/km) in the northeast, to 0.7° (12 m/km) in the Wembley Field area, to 1.2° (20 m/km) in the southern portion of the study area just north of the axis of the Western Canada Basin. Although the Triassic sediments thickened dramatically in the more rapidly subsiding parts of the Western Canada Basin (Fig. 4), the present distribution of the Triassic deposits does not directly parallel the axial plane of the foreland basin. This is due to the truncation of these sediments during the development of the pre-Nordegg Formation unconformity along the eastern margin of the basin.

Paleolatitude

During the Early and Middle Triassic, the paleolatitude of the study area was approximately 30° north of the paleoequator (Habicht, 1979). This commonly accepted configuration is shown in Figure 5 with a pole position established from paleomagnetic data for all North America.

Paleoclimate

Today, depositional indicators similar to those associated with the Halfway and Charlie Lake formations, such as algal-laminated dolomites, anhydrites, and red beds, accumulate in warm (subtropical to tropical) arid climates at latitudes lower than 35° north or south of the equator (Earth Science Curriculum Project, 1967). Thus, these gross climatic indicators are generally consistent with the 30° north paleolatitude derived from paleomagnetic data.

Comparisons with modern environments indicate that the study area probably received less than 25 cm of annual precipitation and may have had annual evaporation rates as high as 2 m during the Middle Triassic. Thus, in addition to having very sparse vegetation, the area would

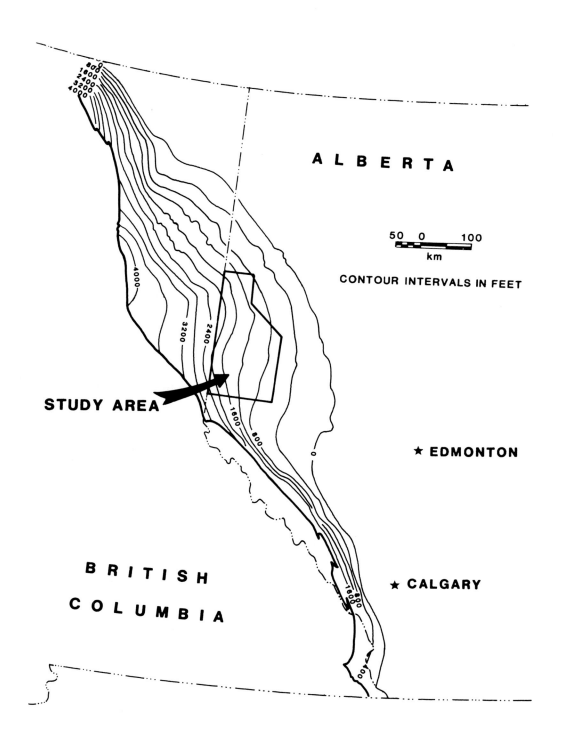

TOTAL TRIASSIC THICKNESS

Figure 4. Total Triassic thickness for the Western Canada Basin.

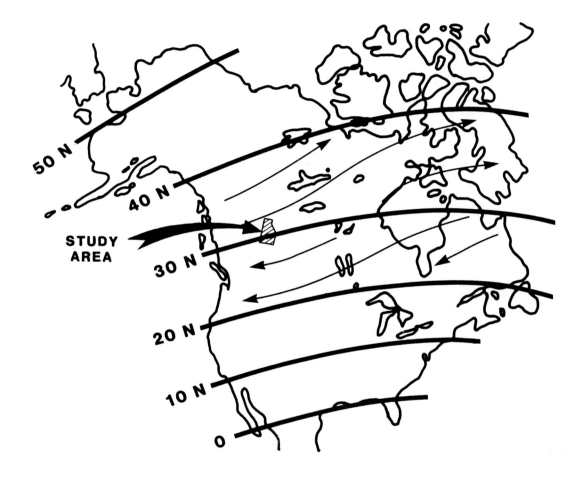

TRIASSIC PALEOLATITUDE AND SURFICIAL WIND CIRCULATION

Figure 5. Triassic paleolatitude and surficial winds.

have had few permanent streams carrying sediments into the basin. There appears to have been only one major zone of sediment input into the Western Canada Basin within the study area in northwestern British Columbia.

Hydrographic Regime

The two physical parameters most important in a paleooceanographic reconstruction are tides and winds. The tides are determined by the configuration of the coastline and by access to the world's oceans. Basin geometry, the embayed Halfway shoreline trends, and facies configuration suggest a high microtidal to mesotidal setting. The paleowind directions are important because they determine the direction and intensity of wave approach to the coastlines within the depositional basin. Paleowinds must be determined after the paleolatitudes have been established.

Above 30° north latitude, the wind blows from west to east (westerlies). Therefore, in the paleolatitude reconstruction shown in Figure 5, the Triassic winds in the study area (29°-31° north latitude) would have been dominantly from the west or southwest. The onshore westerly winds would have generated a wave climate that would have produced easterly to southeasterly longshore currents along the Middle Triassic shoreline. Such currents would carry sediment downdrift to the east and southeast from the areas of sediment influx. A grain size plot for the Halfway sandstone supports this interpretation (Fig. 6).

General Paleogeographic Reconstruction

Halfway shorelines throughout the study area indicate a northwest-southeast depositional strike. This configuration of depositional

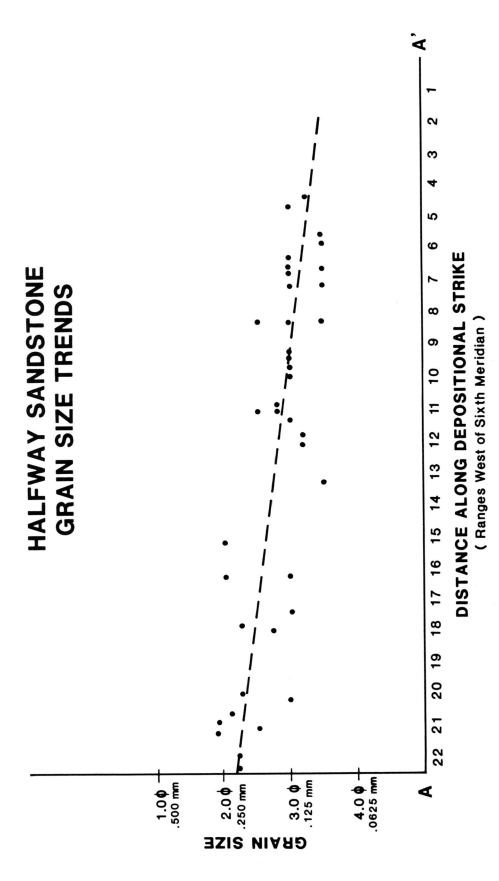

Figure 6. Halfway sandstone grain-size trends.

strike parallels the structural strike of the present-day Triassic basin (Fig. 7). This suggests that the modern Triassic structural basin has slightly accentuated the shape of the original depositional basin and that the shape of the basin strongly controlled the shoreline trends of the Middle Triassic deposits. The northwest-southeast shoreline trend through the study area indicates that the major provenances of the Middle Triassic deposits were the low-relief Paleozoic terrigenous clastic and carbonate sedimentary rocks to the north and east.

Facies Assemblages

The Halfway Formation in Alberta can be divided into distinct facies assemblages composing a barrier/strand-plain sequence:

1. Barrier island facies assemblage
2. Inlet facies assemblage

Both of these assemblages are composed of a distinct group of lithologies easily recognizable in core and in well-log signatures. Characteristic cores and logs from three wells are described below to illustrate the major lithologies. Distribution of the two major assemblages is shown in Figure 8.

Barrier Island Facies Assemblages

The prograding barrier island facies exhibit a characteristic coarsening-upward sequence reflected by the petrophysical log signature (Fig. 9). This facies is elongate and parallel to depositional strike (northwest-southeast), and laterally discontinuous along the direction of depositional dip (northeast-southwest). Where sediment influx into the depositional basin exceeded rates of subsidence, these strand-plain deposits prograded into the basin (south-southwest).

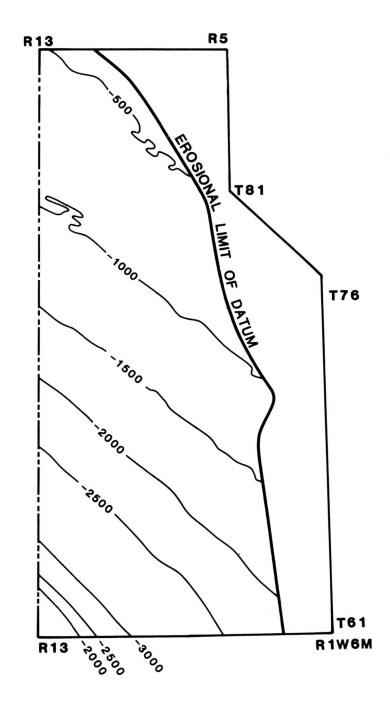

Figure 7. Structure map of datum in Charlie Lake Formation.

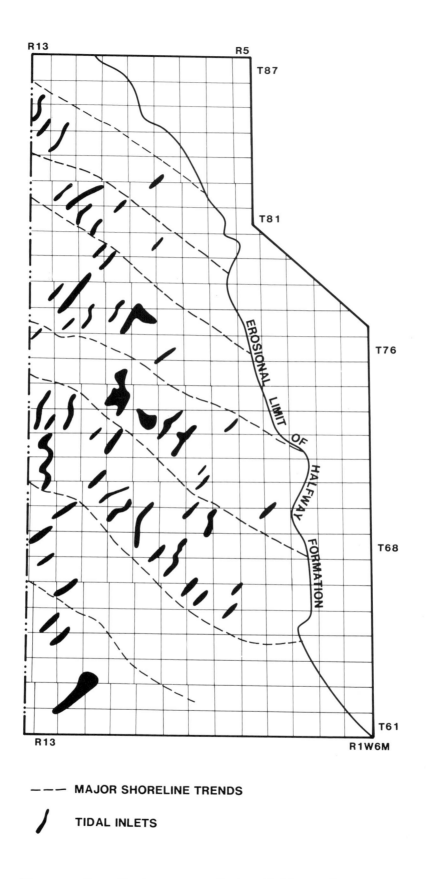

Figure 8. Facies map of the Halfway Formation.

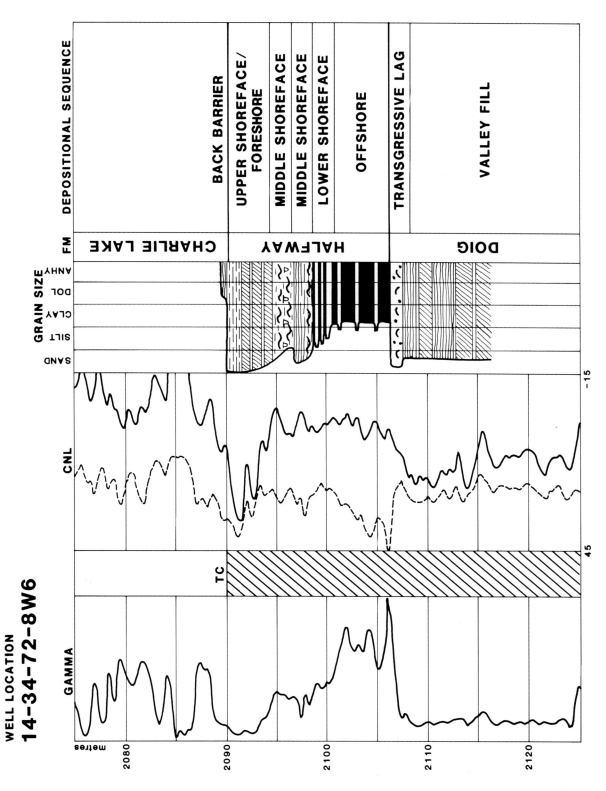

Figure 9. Representative geophysical log signature and graphic core description for barrier island assemblage in well 14-34-72-8W6.

431

The base of the Halfway sequence in well 14-34-72-8W6 lies at the top of the transgressive lag deposit that marks the erosional surface dividing the Halfway and Doig formations. This deposit is composed of abraded shell material, pebbles, and rip-up clasts (Fig. 10A). The deposit grades rapidly upward into the offshore siltstone and shales of the Halfway. The offshore equivalent of the Halfway barrier island facies generally is laminated with minor burrowing (Fig. 10B). Occasional coarser lenses interrupt this fine-grained sequence, and we interpret such lenses to represent storm deposits.

Approximately 6 m above the Doig Formation contact is a gradational change to coarser-grained siltstones and fine-grained sandstones of the lower shoreface. The lower shoreface sediments are planar to wavy bedded and show greater bioturbation (Fig. 10C) than do the offshore deposits.

These lower shoreface sediments grade upward into the planar-bedded sandstones of the middle shoreface (Fig. 10D). The middle shoreface sequence is abruptly truncated at the top by more lower shoreface deposits. We interpret this truncation and stacking to represent a minor transgressive event in the overall regressive Halfway shoreline deposits. The overlying middle shoreface is composed of wavy-bedded fine-grained sandstone and siltstones, and shows evidence of bioturbation (Fig. 11A).

The upper shoreface/foreshore in this well is composed of cross-bedded (Fig. 11B) to planar-bedded (Fig. 11C), medium-grained sandstone. The sandstone exhibits differential and poikilotopic anhydrite cementation patterns. This early anhydrite cementation is responsible for the reduction, and in many cases total occlusion, of

Figure 10. Typical lithologies of the Doig and Halfway deposits in well 14-34-72-8W6. [A] Photo of Doig Formation transgressive lag containing rip-ups, and shell debris in a quartz arenite matrix. [B] Photo of typical offshore marine-laminated siltstones and shales of the Halfway Formation. [C] Photo of moderately bioturbated laminated lower-shoreface siltstones and sandstones of the Halfway Formation. [D] Photo of planar-bedded sandstones of the middle shoreface of the Halfway Formation.

Figure 11. Typical lithologies of the Halfway barrier deposits in well 14-34-72-8W6. [A] Photo of middle shoreface fine-grained wavy-bedded sandstones. [B] Photo of upper shoreface/-foreshore cross-bedded sandstones. [C] Photo of upper shoreface/foreshore planar-bedded sandstones. [D] Photomicrograph showing occlusion of porosity due to anhydrite cementation in the upper shoreface/foreshore sediments.

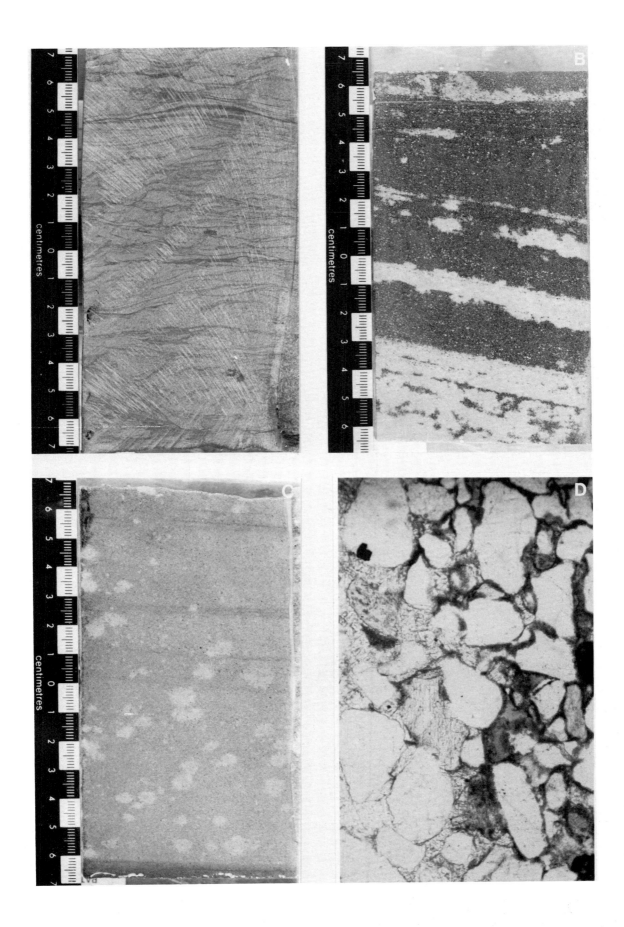

porosity (Fig. 11D) in the barrier island facies. As a result of these anhydrite cements, most of the barrier island deposits are not prospective hydrocarbon reservoirs. The anhydrite cement tends to decrease lower in the stratigraphic section so that the thicker sandstone sequences (inlets) show less occlusion.

The upper surface of the foreshore sequence is capped by a bedded dolomite unit. Some cores show evidence of algal laminations within the dolomite. This dolomite marks the base of the overlying Charlie Lake Formation.

Inlet Facies Assemblage

The inlet facies assemblages in the Halfway Formation are channel-form sandstones and shell-hash conglomerates 0.5-3.0 km wide and 10-25 m thick. They are most continuous along depositional dip (northeast-southwest), but seldom can be traced farther than 3-8 km.

Two cores, from wells 7-14-73-6W6 and 7-5-73-8W6, have been chosen to illustrate the general characteristics of the inlet facies assemblage (Figs. 12 and 13). The core from well 7-14-73-6W6 is an example of an inlet fill composed almost entirely of shell-hash conglomerates. In contrasts, inlet-fill sediments from well 7-5-73-8W6 exhibit an interbedding of clean sandstones with shell-hash conglomerates. Some inlet-fill deposits are in the area of clean sandstones with no shell material.

Both examples of inlet facies lie on scoured surfaces eroded into fine-grained offshore sediments. The bases of both examples contain a pebble and shell lag deposit derived from the underlying sediment (Fig. 14A). In well 7-5-73-8W6, this lag grades upward into relatively clean, cross-bedded, medium-grained sandstone (Fig. 14B). In another core

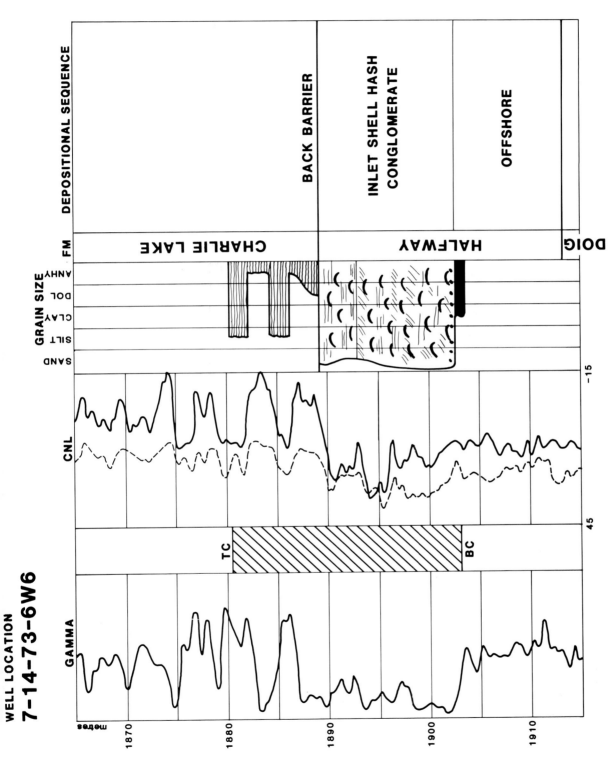

Figure 12. Representative geophysical log signature and graphic core description for inlet facies assemblage in well 7-14-73-6W6.

439

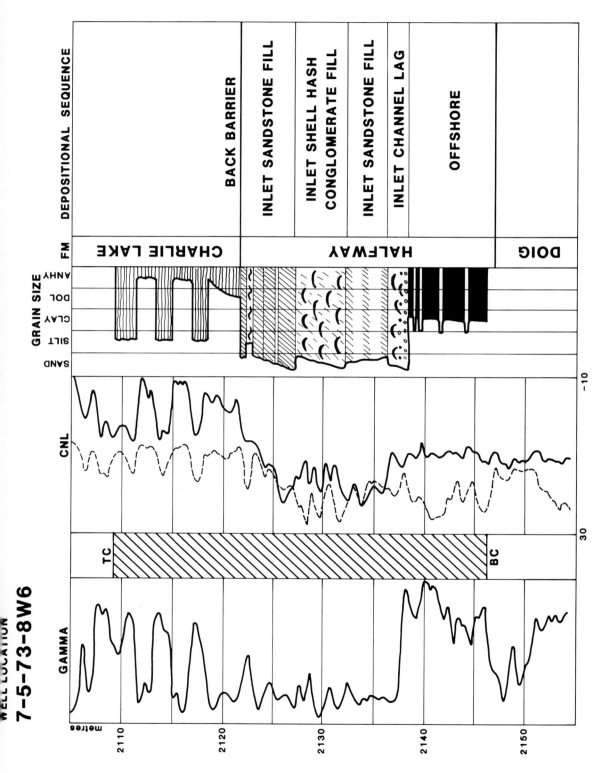

Figure 13.   Representative geophysical log signature and graphic core description for inlet facies assemblage in well 7-5-73-8W6.

Figure 14. Typical lithologies of the inlet deposits in well 7-14-73-6W6 and 7-5-73-8W6. [A] Photo of shale pebble and shell lag at the base of the inlet deposit in well 7-14-73-6W6. [B] Photo of cross-bedded sandstone from the inlet deposit in well 7-14-73-6W6. [C] Photo of cross-bedded shell-hash conglomerates from inlet deposits in well 7-9-73-8W6. [D] Photo of cross-bedded to planar-bedded, granule-rich sandstone at the top of the inlet fill in well 7-5-73-8W6.

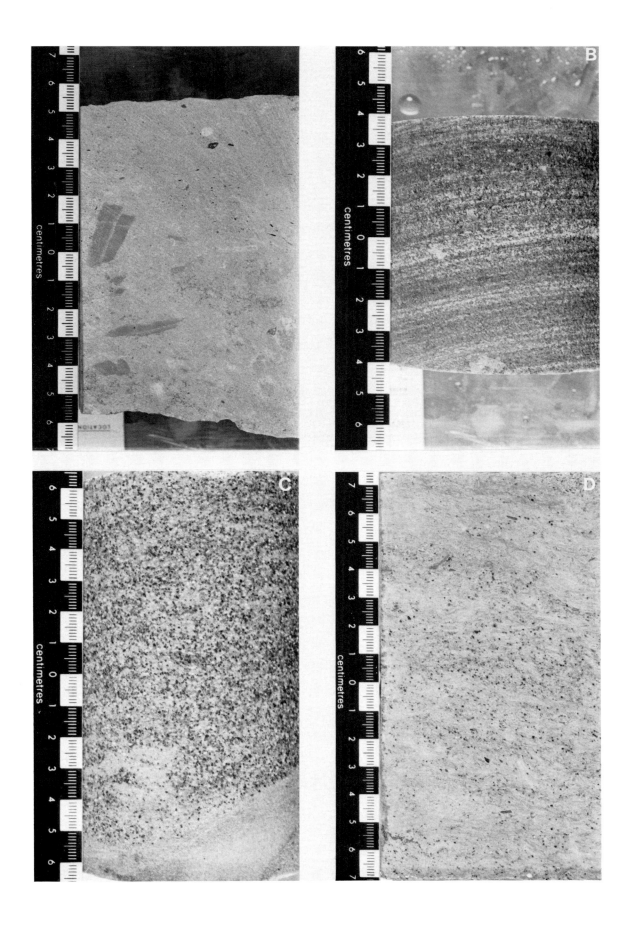

from well 7-14-73-6W6, the lag grades upward into coarse-grained, cross-bedded, shell-hash conglomerates (Fig. 14C). This shell-hash conglomerate comprises the entire 12 m of the inlet fill of this location, except for the upper few meters, which contain some discrete clean sandstone beds. This sequence is capped by an interbedded dolomite and anhydrite unit of the Charlie Lake Formation.

The inlet facies from well 7-5-73-8W6 shows more variation in lithology than do the example from well 7-14-73-6W6. It contains several units alternately composed of shell-hash conglomerates and cross-bedded sandstones. The units appear to be separated by scour surfaces, which may represent multiple stacking of the inlet deposits caused by inlet migration. Such stacking appears in a model proposed by Tye (1981) for Price Inlet, South Carolina. The uppermost meter of the Halfway deposits in well 7-5-73-8W6 contains a planar- to cross-bedded granule-rich sandstone (Fig. 14D), which may represent a prograding recurved spit associated with inlet migration.

Moderate intergranular and moldic porosities may develop in the inlet facies. The moldic porosity has developed as a result of leaching of the shell material in the shell-hash conglomerates (Fig. 15). Its subsequent preservation is due to the sequence of diagenetic events that affected the Halfway sediments. The shell material appears to have undergone early dolomitization that affected the outer rim of the bioclast and created a stable framework. The early dolomitization was followed by leaching, which removed the undolomitized interiors of the shells. As a result of the early dolomitization and possibly early cementation, the moldic porosity was not destroyed by later compaction. Inlet facies form the major hydrocarbon reservoirs within the Halfway Formation in

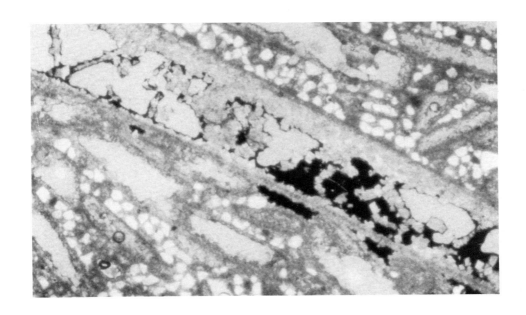

Figure 15. Photomicrograph showing dolomite rims and moldic porosity
developed in shell-hash conglomerates.

Alberta. This is primarily due to the preservation of original inter-granular porosity in this facies and the presence of moldic porosity developed in the shell-hash conglomerates.

Modern Analogues

The lateral distribution of depositional facies encountered in this study can be better understood when compared to similar modern depositional environments. The analogues presented here are each geomorphically similar, which enhances the overall description of the facies examined and depicted in this paper.

Regional Overview

The data suggest that the Halfway Formation was deposited as a barrier-strand plain seaward of hypersaline lagoons and sabkhas backed by a rather low, arid mainland (Fig. 16). Sandstone isopach maps of the Halfway Formation, derived from core and well-log data, suggest that the barrier islands were rather short and separated by numerous, closely spaced tidal inlets at the apex of an embayed coast in western Alberta. The barriers became progressively longer and had fewer, more widely-spaced inlets, away from the apex. The study area probably experienced an upper microtidal to lower mesotidal tidal range (2-3 m) and moderate wave energy, a common depositional setting for modern embayed shorelines.

The shoreline boundaries mark a regional trend around a very wide, funnel-shaped embayment. Such a geometry is indicated by the changes in the geomorphology of the barrier islands farther away from the apex. Hayes and Horne (1982) showed that within a large bight or embayed area, the tidal range is controlled, so that the tidal height

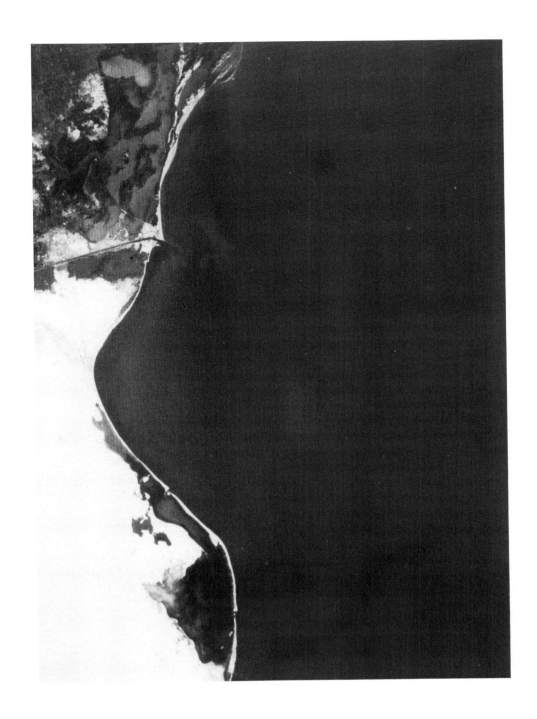

Figure 16. Landsat photo of the Bardawil Peninsula.

increases as the tidal wave proceeds toward the apex of the embayment. As a result, long, narrow microtidal barriers (Davies, 1964) develop on the outer flanks of the embayment, and shorter, often wider mesotidal barriers (Hayes, 1979) form near the head of the embayment.

In this setting, tidal inlets act as conduits between the open sea and backbarrier environments. In mesotidal areas, the inlets tend to be more frequent, display ebb-dominance, and have well-developed ebb-tidal deltas. In the arid climate of Middle Triassic time, the back-barrier environments were probably shallow lagoons, wind tidal flats, and sabkhas containing algal dolomites, as well as the nodular and bedded anhydrites observed in the Halfway Formation cores examined in this study.

Modern Arid Coastline Analogue

The Bardawil Pennisula contains modern environments that may be analogues to the ancient areas that formed the Halfway Formation (Fig. 17). Located along the northern shoreline of the Sinai Desert, this area experiences a mean annual rainfall of 18 cm, a maximum monthly average temperature of 30°C, and an evaporation rate of approximately 2 m per year (West et al., 1979). It is located downdrift of the Nile Delta, which serves as its principal sediment source. Because it is a long, thin microtidal barrier and is interrupted by few inlets, it closely approximates the size and scale of many of the Halfway barriers. The Bardawil Peninsula has prograded eastward under the influence of easterly longshore sediment transport away from the Nile Delta (Levy, 1980). Presumably (as in the case of Halfway Formation), the peninsula is underlain by inlet-fill sequences. This is a result of inlet migration and spit accretion toward the east. The entire peninsula and

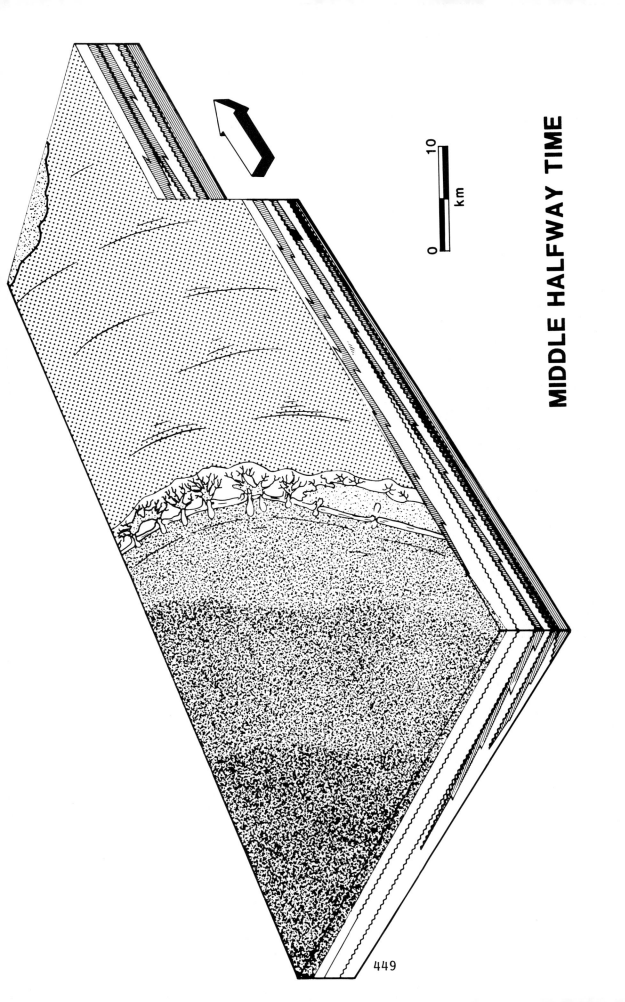

MIDDLE HALFWAY TIME

0 ━━━ 10
km

Figure 17. Paleogeographic reconstruction of the study area during deposition of the Halfway Formation.

449

lagoon/sabkha is underlain by a transgressive lag generated during the Holocene rise of sea level. Thus, a hypothetical stratigraphic sequence of the Bardawil Peninsula area would be (from the base up):

1. Transgressive lag – analogous to the Doig transgressive lag

2. Barrier island and inlet-fill – analogous to the Halfway Formation

3. Lagoon/sabkha deposits – analogous to the Charlie Lake Formation

The modern Bardawil Peninsula sequence is very similar to that of the Triassic Halfway Formation. The major difference in the sequences is the spacing of the inlets. In the Halfway Formation the inlet spacing suggests a high microtidal to mesotidal range, whereas that of the Bardawil Peninsula is microtidal.

Tidal Inlets and Associated Features

Tidal inlet facies make up a significant portion of most modern barrier island complexes and are the most important reservoir-quality facies within the Halfway Formation. Coring programs in modern environments (Pierce and Colquhoun, 1970; Kumar, 1973; Kumar and Sanders, 1974, 1976; Moslow and Heron, 1978) show that inlet-fill sequences have high preservation potential. In the generally transgressive Outer Banks of North Carolina, Tye and Moslow (1983) found that as much as 30% of the underlying deposits consisted of inlet-fill sequences.

In a stratigraphic context, the relative size and continuity of the tidal-inlet sequences in the Halfway are important. Examination of the facies map of the Halfway sandstone (Fig. 8) reveals that the inlet-fill sequences account for a significant proportion of preserved sediments.

At first glance, it would appear that the inlets were extremely wide with respect to the barriers. However, given the predicted hydrographic regime, this is unlikely. An explanation for the width of the inlet fill can be found in modern coring studies. Tye (1981) and Tye and Moslow (1983) show that inlet migration can dramatically control the preservation of inlet-fill facies. Figure 18 shows inlet migration in general terms for a period of about 300 years for Price Inlet, South Carolina. Figure 19 illustrates the extent of inlet-fill sequences actually preserved at Dewees and Capers Islands as a result of inlet migration (Tye, 1981).

Additionally, tidal inlets developing during a rise in sea level tend to occupy topographic lows generated during an earlier sea level low stand (Tye, 1981; Tye and Moslow, 1983; Domeracki, 1982). The topographic lows may be developed by compaction of the fine-grained deposits filling earlier fluvial or tidal inlet scours developed during a sea level low stand. Inlets developed during the subsequent sea level rise tend to occupy these topographic lows and concentrate their deposits there. This occurs in several areas of the Halfway where Halfway inlets occupy the earlier fluvial valleys of the Doig Formation (Fig. 20).

In other areas, the inlets appear to have offset the underlying Doig Formation valley-fill sequences where the valley fills consist of less compactible coarse-grained material. This is the case in the Wembley Field area where two inlet fills flank an earlier Doig Formation channel sequence (Fig. 21). At Wembley Field, the inlets formed the major reservoir units, whereas the shoreline facies generally lack porosity.

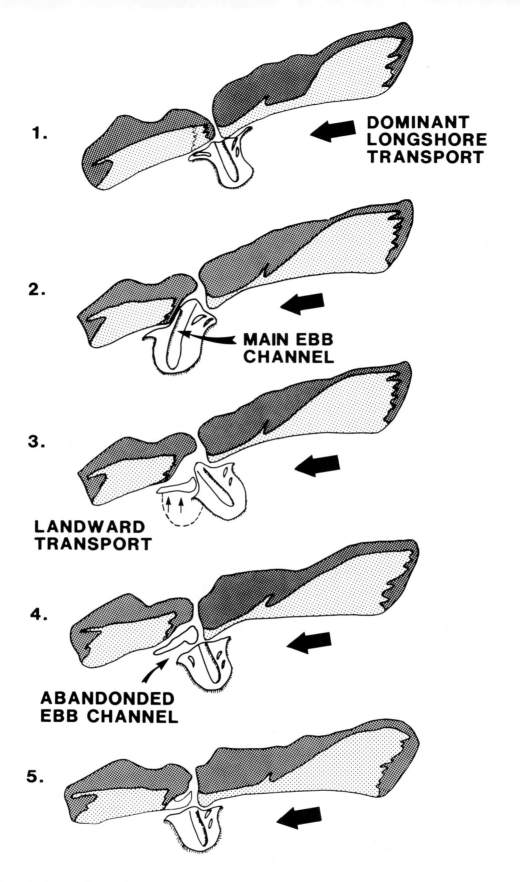

Figure 18. Inlet migration during a 300-year period in Price Inlet, South Carolina (modified after Fitzgerald et al., 1978; Tye, 1981, 1984).

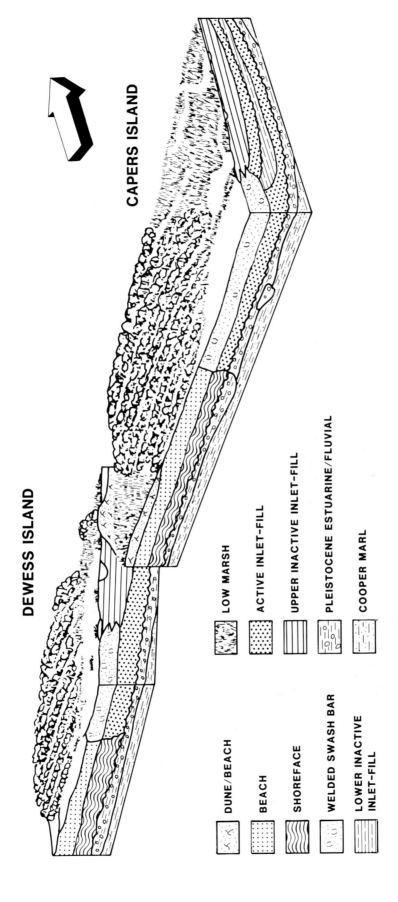

DEWESS ISLAND

CAPERS ISLAND

DUNE/BEACH

BEACH

SHOREFACE

WELDED SWASH BAR

LOWER INACTIVE
INLET-FILL

LOW MARSH

ACTIVE INLET-FILL

UPPER INACTIVE INLET-FILL

PLEISTOCENE ESTUARINE/FLUVIAL

COOPER MARL

Figure 19.   Block diagram illustrating the inlet-fill sequence interpreted for Dewees and Capers islands, South Carolina (from Tye, 1984).

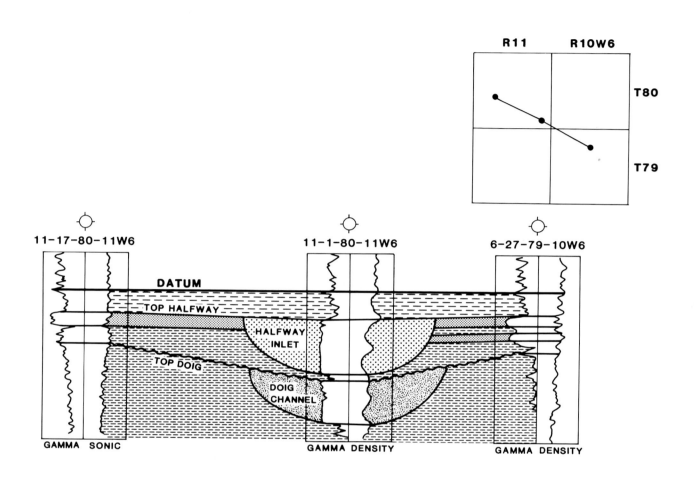

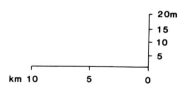

Figure 20. Cross section of Halfway inlet deposits overlying earlier Doig aggradational valley-fill deposits.

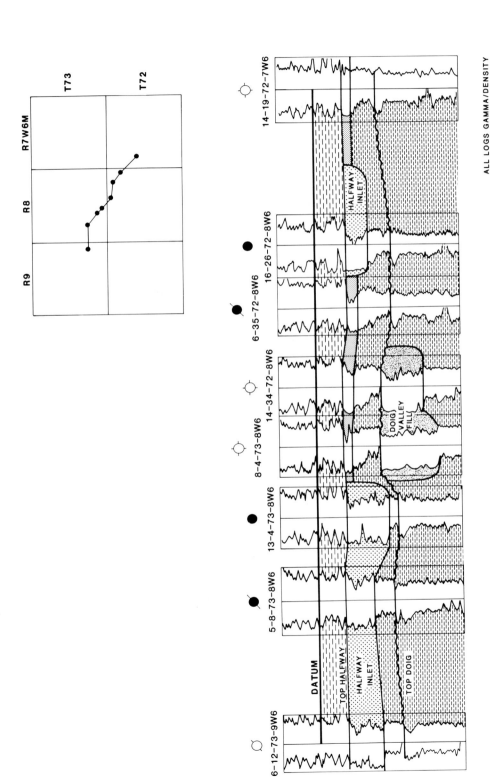

Figure 21.   Cross section through Wembley Field showing two inlets flanking an earlier Doig valley fill.

Conclusions

The Middle Triassic Halfway Formation of Alberta was deposited as a series of barrier islands cut by tidal inlets. The barrier island sandstones trend northwest-southwest and are generally not of reservoir quality. The main reservoir units within the Halfway consist of inlet deposits trending normal to the shoreline that have scoured into the lower shoreface and offshore sediments. These inlet fills consist of shell-hash conglomerates and cross-bedded sandstones that form the major hydrocarbon reservoirs. Moldic and intergranular porosities dominate the reservoir units. The overlying evaporites of the Charlie Lake Formation form an updip seal on the reservoirs. Lateral seals are formed by the tight barrier island deposits and a lateral facies change into impermeable sediments in the backbarrier areas.

References

Armitage, J. H., 1962, Triassic oil and gas occurrences in northeastern British Columbia, Canada: Jour. Alberta Soc. Petrol. Geol., v. 10, p. 35-36.

Barss, D. L., Best, E. W., and Meyers, N., 1964, Chapter 9: in R. G. McCrossan and P. O. Glasster (eds.), Geologic history of western Canada, Alberta Soc. Petrol. Geol., p. 113-37.

Chunta, A., 1969, Halfway Reservoir, PeeJay Field: Calgary Core Conference, Alberta Soc. Petrol. Geol., p. 31-32.

Davies, J. L., 1964, A morphogenic approach to world shorelines: Zeit. fur Geomorph., BD. 8, s. 27-42.

Domeracki, D. D., 1982, Stratigraphy and evolution of the Pawleys Island area, South Carolina: M.S. thesis, Geology Dept., Univ. South Carolina, Columbia.

Earth Science Curriculum Project, 1967, Investigating the earth: Houghton Mifflin, Boston, Mass.

FitzGerald, D. M., Hubbard, D. K. and Nummedal, D., 1978, Shoreline changes associated with tidal inlets along the South Carolina coast: Proc. Coastal Zone '78, v. 3, p. 1973-1994.

Gibson, D. W., 1972, Triassic stratigraphy of the Pine Pass-Smoky River area, Rocky Mountain foothills and front ranges of British Columbia and Alberta: Geol. Surv. Canada, Paper 71-30, 108 p.

Habicht, J. K. A., 1979, Paleoclimate, paleomagnetism, and continental drift: AAPG Studies in Geology No. 9, AAPG, Tulsa, Okla., 34 p.

Hayes, M. O., 1979, Barrier island morphology as a function of tidal and wave regime, in S. P. Leatherman, ed., Barrier islands from the Gulf of St. Lawrence to the Gulf of Mexico: Academic Press, N.Y., p. 1-27.

Hayes, M. O., 1980, General morphology and sediment patterns in tidal inlets: Sediment Geol., v. 26, p. 139-56.

Hayes, M. O., and Horne, J. C., 1982, Variations in depositional systems along shoreline embayments--modern and ancient examples: Research Planning Institute, Inc./Colorado, AAPG 1982 Convention, Can. Soc. Petrol. Geol., Calgary, Alberta.

Hunt, A. A., and Ratcliffe, J. D., 1959, Triassic stratigraphy, Peace River area, Alberta and British Columbia, Canada: Bull. Amer. Assoc. Petrol. Geol., v. 43, p. 563-89.

Kumar, N., 1973, Modern and ancient barrier sediments--new interpretations based on stratal sequences in inlet-infilling sand and on recognition of nearshore storm deposits: New York Acad. Sci., Annals, v. 220, p. 245-340.

Kumar, N., and Sanders, J. E., 1974, Inlet sequences, a vertical succession of sedimentary structures and textures created by the lateral migration of tidal inlets: Sedimentology, v. 21, p. 291-332.

Kumar, N., and Sanders, J. E., 1976, Characteristics of shoreface storm deposits, modern and ancient examples: Jour. Sed. Petrol., v. 46, p. 145-62.

Levy, Y., 1980, Description and mode of formation of the supratidal evaporite facies in northern Sinai coastal plain: Jour. Sed. Petrol., v. 47, no. 1, p. 463-74.

Moslow, T. F., and Heron, S. D., 1978, Relic inlets, preservation and occurrence in the Holocene stratigraphy of southern Core Banks, North Carolina: Jour. Sed. Petrol., v. 48, p. 1275-86.

Peterson, M. S., Rigby, J. K., and Hintze, L. F., 1973, Historical geology of North America: Wm. C. Brown Co. Publ., Dubuque, Iowa, 193 p.

Pierce, J. W., and Colquhoun, D. J., 1970, Holocene evolution of a portion of the North Carolina coast: Geol. Soc. Amer. Bull., v. 81, p. 3694-714.

Tye, R. S., 1981, Holocene barrier-inlet stratigraphy at Capers and Dewees islands, South Carolina: M.S. thesis, Dept. Geol., Univ. South Carolina, Columbia.

Tye, R. S., 1984, Geomorphic evolution and stratigraphy of Price and Capers inlets, South Carolina: Sedimentology, v. 31, p. 665-74.

Tye, R. S., and Moslow, T. F., 1983, Tidal inlet--dominant facies of clastic barrier shorelines: AAPG Bull., v. 67/3, p. 560.

West, I. M., Yehia, Y. A., and Hilmy, M. E., 1979, Primary gypsum nodules in a modern sabkha on the Mediterranean coast of Egypt: Geol., v. 7, p. 354-58.

8 - 30-91

ISBN 0-918985-61-7